Canning

Advanced
Structured Analysis
and Design

PRENTICE HALL SERIES IN SOFTWARE ENGINEERING

Randall W. Jensen, Editor

DEUTSCH   *Software Verification and Validation: Realistic Project Approaches*
DEUTSCH AND WILLIS   *Software Quality Engineering: A Total Technical and Management Approach*
JENSEN AND TONIES   *Software Engineering*
PETERS   *Advanced Structured Analysis and Design*

Advanced
Structured Analysis
and Design

Lawrence Peters

Prentice Hall
Englewood Cliffs, New Jersey 07632

Library of Congress Cataloging-in-Publication Data
PETERS, LAWRENCE J.
  Advanced structured analysis and design.

  Bibliography: p.
  Includes index.
  1. Electronic data processing—Structured techniques.
2. System analysis. 3. System design. I. Title.
QA76.9.S84P48 1988   004.2'1   87-11391
ISBN 0-13-013137-7

Editorial/production supervision and
  interior design: Denise Gannon
Manufacturing buyer: Richard Washburn

© 1987 by Prentice-Hall
A Division of Simon & Schuster, Inc.
Englewood Cliffs, New Jersey 07632

Printed in the United States of America

10   9   8   7   6   5   4   3   2   1

ISBN 0-13-013137-7 025

PRENTICE-HALL INTERNATIONAL (UK) LIMITED, *London*
PRENTICE-HALL OF AUSTRALIA PTY. LIMITED, *Sydney*
PRENTICE-HALL CANADA INC., *Toronto*
PRENTICE-HALL HISPANOAMERICANA, S.A., *Mexico*
PRENTICE-HALL OF INDIA PRIVATE LIMITED, *New Delhi*
PRENTICE-HALL OF JAPAN, INC., *Tokyo*
PRENTICE-HALL OF SOUTHEAST ASIA PTE. LTD., *Singapore*
EDITORA PRENTICE-HALL DO BRASIL, LTDA., *Rio de Janeiro*

To my daughter Jean and my son John
in the hope that this book
might pay for their college education

Contents

Preface

Nearly a decade ago, it became obvious that a growing number of software applications required new problem solving techniques. The early works on these new methods really addressed the first wave of automation—that is, they provided very effective means of assisting in the automation of what were formerly manual tasks. At this point in time, we are in what is, at the least, the second wave. We are attempting to replace some applications and integrate others in a cohesive system. Combining this with the timing and database aspects of *today's* software engineering environment we obtain an environment beyond what these methods were directed at. Hence, the necessity for this text.

Structured Analysis and Design are the most recent software development phases to receive the attention of conferences, books, and papers. For whatever reason, once the software industry had "discovered" structured programming, everything else had to be structured. Many of the newer analysis and design methods available today use one or more of the "magic" words—structured, advanced, analysis, design, information, systems, modeling, or methodology—in combination or in conjunction with other terms. Readers encountering those words in the title of this book may wonder, therefore, exactly what we will be discussing here.

This book presents a highly modified form of Structured Analysis and Design, which is the result of an evolutionary process. The form of these methods did not come about instantly nor are they the result of speculation. They are the result of direct use and consultation with those who have attempted to apply the concepts and principles set forth in the books, *Structured Analysis and System Specification* by T.

DeMarco, *Structured Design* by E. Yourdon and L. Constantine, and *The Practical Guide to Structured Systems Development* by M. Page–Jones. The preceding works represent a response to problems that existed a decade ago. The power and broad scope of applicability of them is evidenced by their continued widespread use. Just as times change, problems and approaches to solutions change.

What makes this text *Advanced* is that it integrates several topics that are usually treated separately, and modifies the methods based on real experiences with many diverse projects, to meet today's needs. The use of Structured Analysis, Structured Design, and Information Modeling are treated as a *continuous process*. They are not treated as easily distinguishable steps with identifiable starting and ending points. The orientation is toward what does and has worked, *not* what is ideal. The notation employed here is a departure from some earlier approaches. In developing it, attention was paid to a growing body of experience regarding complexity and communication. A conscious decision was made in developing this *Lifecycle-Based Software Engineering Method* that graphics should be simple, focused, and contain what can *best* be represented graphically.

Finally, one might ask, "Why would anyone in their right mind want to write a book that attempts to combine and extend what is presented in other works?" I am not sure that I can answer that but I can say that creating this text has been a true learning experience. For example, the publication of this book was delayed by more than two years due to revisions necessitated by problems I encountered in my consulting practice analysis as well as design problems that I had never seen before. I am certain that each reader of this text has encountered (or will encounter) problems that are not addressed here. This can happen, despite our best efforts, if for no other reason than the amount of time it takes to get a book into print. It is my hope that enough material, guidance, and examples have been provided so that further expansions can be made by readers on an extemporaneous basis. Thank you for your trust in purchasing this book. I look forward to hearing from you with suggestions for revisions for its next edition. Enjoy!

Kent, Washington   Larry Peters

Acknowledgments

Writing a book does not have to be a thankless job. There are many people and organizations whose insightful comments, specialized problems, and imaginative approaches have greatly enriched this author's experience and the content of this text. Specifically,

Harry Walker, whose ability to describe complex issues in information systems concepts through simple, realistic examples and not become annoyed at inept questions, is a gift.

Bernard Goodwin of Prentice-Hall whose patience and understanding have been stressed to new limits.

Randy Jensen, whose invitation to write "about 75 pages of text on Structured Analysis and Design," started this book.

My students and consulting clients who were instrumental in the creation of a more "terrain-oriented" version of these methods in response to real problems.

Mssrs. Hewlett and Packard whose HP-110 portable computer's ability to go anywhere and work anywhere is a primary reason why this book ever came into being.

The pilots and navigators of Delta, American, United, and Alaska airlines who kept my flights smooth enough that I was able to create about three-quarters of this book while airborne.

Karen Anderson and Carol Wright for entering text and updates and not being afraid to declare that something did not make any sense.

Finally, to my wife, Cathleen, whose insistence that this book be finished was a key factor in its completion.

Using This Text

This book is organized into three sections. They are intended to be read in order:

- The first section describes the viewpoint upon which this text is based. It relates the three disciplines (i.e., process modeling, information modeling, and event modeling), describing the advantages and shortcomings of each.
- The second section describes the Structured Analysis method, its notation, use, and the three types of modeling approaches that are now brought together in it.
- The third section describes Structured Design, its relationship to Structured Analysis, and its implementation.

The order in which these topical areas are discussed needs some explanation. Chronologically, Structured Design was the first of these methods to be published. It reflected a shift in thinking away from the importance of efficient, error-free code to increased importance of system architecture—relationships between/among modules. This was followed a few years later by an increased sensitivity to issues revolving around the specification of systems. Structured Analysis was a result. Basically, the trend was toward how to specify a system in such a way as to aid the effort in design—particularly, Structured Design. This was followed a few years later by Information Modeling, which reflected increased concern over information definition, description, and policy issues.

This entire evolutionary process has been *away* from code and *toward* data.

Stated another way, we have migrated *away* from a process orientation and *toward* an information orientation. This shift is only partially reflected in this book's organization. We present Structured Analysis first primarily because it will probably be of greater interest to more readers than Information Modeling, which is presented second. Information Modeling can have a significant impact on both Structured Analysis and Structured Design. The last section discusses Structured Design, which is the culmination of the other two activities.

A cross-reference of chapters and generic job descriptions is presented below, showing the level of importance or utility to each job type. This is strictly an advisory, offering some guidance to those who may wish to concentrate their reading/learning time on the topics most directly related to their current work.

							CHAPTER								
ROLE/POSITION	1	2	3	4	5	6	7	8	9	10	11	12	13	14	15
Systems development manager	X	X			X			X	X	X					
User/client/sponsor	X	X	X		X	X		X	X	X	X				
Systems analyst	X	X	X	X	X	X	X	X							
Systems designer	X	X	X					X	X	X	X	X	X	X	X
Programmer/analyst	X	X	X	X	X	X	X	X	X	X	X	X	X	X	X
Software engineer	X	X	X	X	X	X	X	X	X	X	X	X	X	X	X

Definitions of these roles as intended here are presented below:

Systems development manager. One who directs (or at least tries to direct) the efforts of others who are developing and maintaining systems. This person is held responsible for the results of such efforts on the development side of the organization.

User/client/sponsor. One who contracts for the development (or modification) of systems. This person may also be an eventual user of the system. This role includes those who act as focal point on behalf of users.

Systems analyst. One who may play a combination of roles throughout the systems lifecycle. This person may perform in any or all of the roles listed here.

Systems designer. One who is responsible for the creation of a system architecture and detailed system blueprint (in our case, software) which meets the specification set forth during the analysis activity.

Programmer/analyst. One who may play a combination of roles throughout the systems lifecycle. This person may perform in any or all of the roles listed here.

Software engineer. One who is responsible for the definition, specification, design, and detailing of a software system but not necessarily its instantiation into code.

As noted above, software personnel in many organizations end up performing in several, if not all, of the roles described above.

Throughout this book we present the analysis and design of systems as a *process* rather than as distinct, quantizable activities. Even so, if this book is used as a class textbook, it can be covered in two or possibly three units (quarters or semesters).

PROLOGUE

The book's organization requires some explanation. It begins by discussing various models of what happens during the software systems and design phases. This Lifecycle View is carried on into the later sections of the text, which deal with the "nuts and bolts" of how to actually do Structured Analysis and Structured Design.

We have expanded the notion of what is appropriate during the analysis phase and included what some may feel are diverse elements. The general topical area of Information Modeling has been incorporated into Structured Analysis and Structured Design. Hence, we do not treat it as a separate activity but as one which is intrinsic to analysis and design.

The entire process of analysis and design is treated from the standpoint of the development and refinement of a series of models. Each model is based on its predecessors, and we build upon what was previously done rather than starting anew. One model represents the processes being examined, the data that they operate on, the data they output, where the data come from and go to. We will refer to this model as the Process Model. Another model, which may evolve as part of the development process, describes in detail the data itself—that is, how data relate to other data, what the policies regarding the data are, what specific type each data element is. We have recently begun to accept the fact that of the two, code and data, data is far more important and less easily replaced than code. Hence, it is of far more importance to the enterprise. This model we will refer to as the Information Model.

These two models alone describe all of what is contained in the two primary works describing Structured Analysis and Design. In fact, all of the works that describe one method or another in the field of software system specification and design are directed at process models. The integration of the Information Model into a methodology has not occurred prior to this text.

But an additional type of system characteristic is important and is usually not modeled. This characteristic is related to the response of the system to the various types of stimuli that it can encounter, such as user requests in the form of interrupts, device timeouts, and actions by the user or outside systems. These can be described using a modeling approach we shall refer to as the Event Model.

These three models form a complete set, more complete than any of the individual methods developed thus far. We will discuss all three types of models in both the analysis and design phases.

The models' form and content change as we move from the specification phase through design. For example, the process model in the analysis phase takes the form of dataflow diagrams (input-process-output oriented charts) and supporting material describing data policy, transformation, and definition. In the design phase, the Process Model takes the form of hierarchically organized "blueprints" of the system depicting the population of modules, their control relationships, and the information they exchange. The form the model takes in the design phase is derived from what was developed during the analysis phase. This evolutionary feature is unique to Structured Analysis and Structured Design and ensures that there will be a strong correlation between what was specified and what was built. Similar comments apply to the Information and Event Models.

This book describes the development of each of these three models. The emphasis is on their interrelationships. The means of derivation of one model from another is of particular importance and is highlighted throughout. The models

TYPE OF MODEL	DEVELOPMENT PHASE		
	STRUCTURED ANALYSIS	STRUCTURED DESIGN	IMPLEMENTATION
PROCESS	Dataflow Diagrams Dictionary[1] Pseudocode	Structure Chart Dictionary[2] Pseudocode[4]	Code Dictionary[3]
INFORMATION	Entity-Relationship Diagram Dictionary[1]	Database Design Dictionary[2]	Implemented Database Dictionary[3]
EVENT	Event Model[5] Dictionary[1]	Event Model[6] Dictionary[2]	Queuing Model Dictionary[3]

[1] This dictionary includes data definitions: pseudocode for each process in the analysis model; descriptions, attributes, and content for the entities and relationships in the E-R diagrams; and descriptions of the events and states in the Event Model.

[2] This dictionary contains everything found in the analysis version of the dictionary plus definitions of all flags and pseudocode for all modules that were added as part of the design process or not otherwise present in the analysis.

[3] This dictionary has been updated to incorporate all information developed during the design activity which has been modified as a result of implementation considerations.

[4] The pseudocode referred to here incorporates any and all flags employed in the Structure Chart. All pseudocode is contained in the dictionary as well. Much of this may be identical to or based on the pseudocode that was developed during the analysis.

[5] Event Model refers to a simplified form of state transition diagram.

[6] This Event Model would be further refined and leveled based on design "discoveries."

Figure P–1: Relationship of Phases to Tools and Model Types
(Extracted from the seminar, "Structured Analysis for Real-Time Systems," by Software Consultants International, Ltd.; Kent, Washington; Copyright 1988. Reprinted by permission.)

described also include supplemental methods that meet needs not originally addressed in Structured Analysis and Structured Design. The models, their relationship to phases, and the concepts and tools associated with them as they will be presented in this text are portrayed in diagrammatic form in Figure P-1. This diagram will be repeated throughout the text to remind the reader which conceptual tools will be discussed as well as the form they will take.

The need for creating three coherent models of the system is evident if one examines the origins of Structured Analysis and Structured Design. They were developed in the mid-seventies in response to the data processing industry's needs. At that time, most software development efforts were directed at the automation of manual processes and the centralization of formerly independent batch processes. As a result of what was achieved, today's efforts are significantly different. They involve responding to altogether new needs within the business community. These needs have been perceived as a result of increased insight into the nature of various enterprises, allowing the observation of opportunities for a new service to customers. Hence, today's needs and those of the foreseeable future are not easily met by methods requiring that we have a current system to study and gain insights from. Often we know only the kinds of data that the envisioned system will process and the kinds of things it is to do with that data. In such cases, we need to be able to derive the Process Model from the Information Model. Thus, all three views are important, and the ability to derive any one or two of them from the one that we can quickly obtain is a prerequisite for an effective software analysis and design method for the 1980s and beyond. This book is an attempt to deliver that ability.

Advanced
Structured Analysis
and Design

PART I

Structured Systems Overview

Engineers have spent a long time trying to describe systems in effective ways. Early in the software industry's life, we utilized the flowchart because it provided a convenient mechanism for describing what was important to us—control flow. Our concerns revolved about whether or not the code we had developed gave the right answer. As our confidence grew, so did the size of our programs, and we began to see distinct advantages in being able to modularize our programs into systems of interacting, somewhat independent units. For one thing, when errors had to be corrected, we would need only to recompile and relink the changed portions of the system.

Modularity seemed like *the* answer to programming's problems, but soon other problems dominated our thinking. One was the notion of software understandability. Debates raged over the advisability of using the unconditional transfer-of-control or GOTO statements. We needed, however, to do a lot more to a module than just count occurrences of GOTOs to make it viable over its useful life. We became aware that a module's content *and* its relationship to other modules were the two primary keys to the maintainability of a system.

Software was perceived as overly expensive to develop. Replacement of an entire system just because it was programmed in another era was logistically unthinkable and an economic nightmare. Nonetheless, pressures mounted to do away with many of these systems. The increased reliability and lower costs of hardware made the use of real-time applications more feasible. This trend was paralleled by an unprecedented growth in the use of databases. Both of these trends have pushed current software analysis and design methods beyond their scope of applicability.

The shortcoming of analysis and design methods previously published is that they are concerned primarily with the representation of systems as processors of data, ignoring other aspects of *today's* software systems. Such ignorance has proven fatal to many projects. These additional aspects are:

Real-time characteristics, their graphic representation and description

Database requirements, their definition and graphic representation

The systems view expounded here emphasizes the incorporation of several models. This view coordinates and synchronizes the models so that they are consistent with each other and address the factors that are critical to the success of an analysis or design effort. The primary problem this strategy addresses is the fact that software people tend to view the various properties of software systems as being independent. Prior to employing this new strategy, their interrelationships have proven elusive to represent and difficult to cope with.

In this first part of the book, we will propose a way of depicting the interrelationships among these various properties—an integrated view of software systems development which will be expanded throughout the remainder of this text. The emphasis will be on viewing software systems as composed of several different types of properties. Some of these properties are more predominant in some systems than others. For example, in the case of software systems which enable a high-level manager to query a large database for the answers to certain ad hoc questions, the database and information properties of the system are more prevalent than in, say, an embedded tactical software system that is part of a fighter aircraft's electronic countermeasure systems. In this latter system the predominant factor would be the ability of the software to respond to various situations in real time. The volume of data may or may not be a critical factor, depending on the exact nature of the real-time system.

Throughout this text we will emphasize the employment of a systems approach to the practice of analysis and design. In it we attempt to present this portion of the software lifecycle as a discipline rather than an "automagical" (*auto*mated *magic*) process. This part of the text details exactly what Structured Analysis and Structured Design encompass.

CHAPTER 1

Software Lifecycle Models

When we speak of software development activities such as analysis and design, we imply that there exists some model of all possible activities within software development by which some of these activities have been allocated to "analysis" and others to "design." Other aspects of software development and maintenance could be treated in a similar manner. The difficulty is that we need to define what such terms mean. One way to describe these activities and their relationships to one another is to use the concept of Lifecycle—that is, a model of the activities which occur during the lifespan of the system, starting with project initiation and ending with system replacement. In the case of software, we are usually dealing with maintenance (a.k.a. "Land of the Living Dead") as the final phase.

A Software Lifecycle describes the stages the software system goes through from birth to death. Many of us in software engineering feel that this model is of great importance to the overall success of a software project. Using the "wrong" model ensures difficulty, perhaps failure. Using the "right" one ensures at least the possibility for success [1]. How can we tell the right models from the wrong ones? Anyone who researches the topic discovers that several lifecycle models have been proposed, implemented, and refined, and there is no clear-cut indication of which is best.

In this chapter we will examine four classes of lifecycle models, review their inherent assumptions, and relate experiences with them. Our purpose is to provide insight into the role that Structured Analysis and Structured Design play in the lifecycle, where they fit in, and what sorts of lifecycle models seem appropriate for use in conjunction with them.

1.1 DEFINITION OF SOFTWARE LIFECYCLES

The term "Software Lifecycle" may have as many definitions as there are software engineers. However, many of these definitions have much in common. Several basic ones come to mind:

> A Software Lifecycle is a model used to explain and help us understand a software development and maintenance process [S1]
>
> A Software Lifecycle is a step-by-step breakdown of the software development process [S1]
>
> A Software Lifecycle is a list of things to do [S1]
>
> A Software Lifecycle is both a management and technical tool for organizing, planning, scheduling, and controlling the activities associated with a software development and maintenance effort [S1]

Taken together, these basic definitions constitute a reasonable description of the term Software Lifecycle.

The basic descriptions have certain features in common. They describe the Software Lifecycle as

> A model, plan, or guide
>
> Activity and process oriented
>
> Of use to both those managing the work and those doing it

In the remaining sections of this chapter we will examine four basic lifecycle models and introduce the role of the structured methods.

1.2   THE "SLAM DUNK" LIFECYCLE MODEL

Whether we admit it or not, this is probably the most common form of Software Lifecycle in use today. It results from a lack of desire to plan and detail what will be done *before* we actually do it. In the "Slam Dunk" model, we begin coding almost as soon as the project begins. The basic philosophy is that it is easier to write the programs, debug them, and get them to actually do some approximation of what the client wants them to do than to "waste time" applying all this "science." These are the expressed sentiments of several software managers. There are many others who will not say as much but whose behavior indicates agreement with the thrust of such a comment.

Where do such views come from? The following three factors help to foster them.

Psychology.   Studies done over the last ten years [2] indicate that software development people possess certain common and unique psychological characteristics. One of these characteristics, called Growth Need Strength (or GNS), is a strong need for challenge and accomplishment—that is, to have actually *done* something. Coding satisfies this need, because we can see the fruits of our labors in terms of code that migrates from possessing syntax errors to being free of them to compiling error-free to executing error-free. Each of these steps is a stage of accomplishment. Something was actually done. Spending the same amount of time defining the overall system architecture, detailing the requirements, refining some needed definition, or planning our activities does not provide nearly as obvious an example of accomplishment—hence, the tendency to avoid such activities.

Another characteristic of people in software development is a low Social Need Strength (SNS). This combination of low SNS and high GNS was unique among the professions studied. A low SNS means that the person is more capable of working independently, rather than as a part of a team. Perhaps this is why the walkthrough [3], which was publicized almost two decades ago, has not swept the industry.

Education.   The education that most people receive from grade school through the bachelor's degree level in college deals with deterministic problem solving. The problems we are required to solve are well stated; they require no investigation and no "guesswork" or judgment calls. They have one and only one answer and clear-cut completion criteria. A person who comes out of that sort of an environment and goes into software has little motivation to deal with the unknown—*analysis and design.*

The Nature of Software.   Most software tasks look relatively easy until we actually begin work on them. Why? One reason appears to be that software is not physical. No one has ever actually *seen* a computer program. Oh, we have seen computer tapes, diskettes, and listings, but the programs themselves are never seen. This has a lot to do with what we perceive the fabric of a software development task to be. When we examine a physical activity such as the construction of a house, we can, a priori, identify many stages that its construction will have to go through. Further, we can mentally "see" that if changes are made at any stage, probably the subsequent stages will be affected and the overall schedule will have to be lengthened. The construction of software, however, is really outside our first-hand experience. Hence, it appears relatively simple to meet the customer's requirements with no analysis or design, because we cannot even identify with just what is involved in meeting those requirements. The so-called "*N*th-generation" languages have tended to aggravate rather than alleviate this problem, since they make it even easier to prematurely implement and reimplement a system or part of it ad infinitum. In fact, some of these encourage the use of monolithic control structures and discourage the practice of modularity and passage of control to and from a subordinate module and boss module.

Another cause related to the perceived nature of software is psychological. It resides in the ego that software people seem to possess. That is, they believe (egotistically) that if they can conceptualize the system at a high level of abstraction (very little detail involved), then realizing or implementing it will be easy. Thus, the lifecycle consists of "get an idea" followed by "code." Hence, the "vaporware" phenomenon, wherein software products are announced, delayed, declared to be RSN (Real Soon Now), and then are *late*.

The "Slam Dunk" approach, then, is based on the theory that since there will be so many "bugs" to get out of the code, there is precious little time for such scientific stuff as analysis and design. The overall time line associated with this model of software development and maintenance is shown in Figure 1.2-1.

Experience with the "Slam Dunk" approach to software development has not been good. In one way or another, this practice has seriously hurt the credibility of software development people with respect to costs, schedules, and product quality. The structured methods can mitigate these effects, as we shall see later in this chapter.

We should note, too, that the lack of a common education base for software people combined with the limitations of the technology available has enabled us to violate several engineering principles—especially the principle that construction begins only *after* a blueprint has been generated, approved, and accepted. In the "Slam Dunk" lifecycle approach, software developers behave as though systems could be built *without* a blueprint.

BEGIN CODING — — — — — CONTINUE CODING — — — — — QUIT

Project start   Time

Figure 1.2-1:   The Slam Dunk Lifecycle Model Time Line [S1]

1.3 THE BAROQUE APPROACH

A response to the lack of discipline and structure exhibited by the "Slam Dunk" approach is to "decree" discipline. The concept is that each stage of software development will be completed before the next is begun. The stages that this "Baroque" lifecycle dictates are analysis, followed by design, followed by code or implementation. This sounds great—until we actually go out and attempt to use it. What we find is that the process of analysis, for example, is an indeterminate one. That is, we could go on defining and redefining the requirements for the project indefinitely. Why? Because there are always new discoveries about how business is currently being conducted that are intriguing. These almost always lead to other discoveries, which lead to others, and so on.

In decreeing a lockstep arrangement of tasks in order to avoid the ''Slam Dunk'' phenomenon, some software managers have created a worse situation, wherein the project enters the analysis phase and never seems to get out of it. This is analogous to the phenomenon discovered in space called a ''Black Hole.''

Part of the difficulty lies is the fact that the interaction which must take place between the analysis and design activities is disallowed. Also, what constitutes a *complete* analysis is not defined and is nearly impossible to describe. Hence, project personnel are working toward an undefined goal, and they are trying to ignore the fact that as we attempt to define the requirements, they change. One reason for their changing is that, as we try to understand the requirements, we learn more about them and discover other nuances that must be considered. Another is that as the analysis is conducted, the client begins to broaden the scope of inquiry and begins to ''see'' the advantages of automation or more advanced automation in ways that are not predictable. So the client begins to request that the system do more than originally planned for. The whole situation is summed up in Figure 1.3-1.

The Baroque lifecycle model does not work for the simple reason that software development is *not* a deterministic activity. It requires refinements and interactions between lifecycle stages, which the Baroque approach attempts to legislate out of existence. However, the attempt to use it does make some important contributions. One is that it shows the necessity to have a clear-cut definition of just what is to take place during analysis and design. Without it we may enter a phase and never leave. Another contribution is, in showing that the structured methods have a benefit which is often overlooked, they provide a list of things to do or major products to be created during each of these phases.

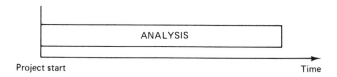

Figure 1.3-1: Timeflow Associated with the Baroque Approach [S1]

1.4 THE WATERFALL APPROACH

A more recent development in the area of software lifecycles is the Waterfall approach [2]. It attempts to correct the shortcomings of the Baroque approach by recognizing an advantage in having interaction between the phases. This interaction is accomplished by having the phases overlap with respect to time. The results of one phase are fed into the next, beginning with the analysis phase. An example of the time line associated with the Waterfall approach is presented in Figure 1.4-1.

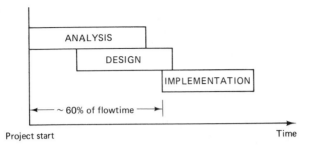

Figure 1.4-1:   Time Line Associated with the Waterfall Approach [S1]

As new and different as the Waterfall approach may have sounded when it was first introduced, it represented what was accepted practice in other engineering fields. These other fields encourage and heavily depend on the kind of interaction the Waterfall approach promotes. However, other fields of engineering are much more disciplined in their use of these interactions and have established controls to prevent project overruns and scheduling difficulties.

As useful as this approach has been, it has some shortcomings. The major one is that in a long-term development effort (say, two or more years from start to finish), it is often more than a year before any working product or model of a product is available for the customer to examine and critique. At this point, because of the time lag, the project may be in serious jeopardy. The observation is based on experience, but the mechanism by which the problem occurs is uncertain. It is proposed that the client's expectations are always growing and changing. If a year or more goes by without any perceivable system being available, then the client's expectations may very well have outgrown what is possible within the constraints of flowtime, personnel, and budget.

1.5 THE PROTOTYPING APPROACH

This approach has shown itself to be a useful adjunct to the Waterfall lifecycle model. Prototyping is a standard practice in nearly all engineering professions. It was "discovered" sometime in the mid-seventies and has prompted a number of articles and textbooks and a series of conferences. Its main advantage is that it can give users an early "feel" for what the system will be like before it is built. Among its disadvantages are that

> The lifecycle and management model deteriorate into what amounts to a "Slam Dunk" development. This is marked by runaway software development, although what is learned is supposed to be fed back into the design; in fact the prototype itself is refined and adopted as the delivered system.

The user may find the prototype so attractive and useful that the development team may be directed to abandon the remainder of the development effort.

The author's first experience with this type of lifecycle may offer some useful insight into the circumstances under which it is appropriate to prototype a system. In the mid-seventies, as part of a computer-aided manufacturing (CAM) software development effort, we were dealing with a client who really did not have a clear picture of what the system we had contracted to build was to do. A few things were known. The system would be supported by a "number-crunching" mainframe and utilize a terminal that had a memory screen. It would be an interactive, real-time system. Granted, memory displays are not appropriate for use with interactive systems, but the client owned them and wished to use them for this purpose. We pointed out that there would be problems for the users of such a system, but the client indicated that these problems were understood and directed us to proceed anyway.

The system would support the use of certain software development methods, but few of the details were agreed upon. As manager of the project, I became concerned that, under the circumstances, it was highly unlikely that the client would be satisfied with whatever was delivered. That is, undefined requirements can never be satisfied. We proposed that we link together the graphics interface and the database manager to be used on this system and implement only a few of the dozens of functions that the final system would support. In accordance with our project plan, this prototype would be delivered about three-fourths of the way through analysis. Owing to the overlap of phases, this corresponded to about the first fourth of the design phase and a very thinly sliced prototype development phase.

Throughout the prototype development, the client was furnished with screen mockups and the command structure. The users the client supported could not wait to get their hands on the prototype. Once they did, things changed dramatically. For instance, they finally realized what sort of interaction problems users would face with a memory display. Being told that there are problems and actually experiencing them are two very different phenomena.

A popular topic related to the subject of prototypes is the concept of "throw-away" code. From both an economic and a methodological standpoint there should be considerable concern about this one. Few customers wish to pay for the development of anything that is to be "thrown away." From a methods and disciplined system-development standpoint, the knowledge that the prototype will be discarded can and has put more than one development team into a "Slam Dunk" frame of mind. Granted, this depends on how prototyping is presented to the team by management and what quality controls are put into place, but the dangers are there.

An alternative way of employing prototyping is to use it as a means of implementing a preselected subset of the modules within a well-defined architecture developed through a design effort. This ensures that at least some of the developed code will *not* be thrown away. At worst, much of it will have to be changed.

Experience to date indicates that prototyping is most appropriate for systems which have a significant interaction with users. Interactive, real-time systems are prime candidates in this regard. Whether prototyping is intentionally used or not, the use of mockups of displays and output reports is a form of prototyping and is advised in any case.

One form of the Prototype lifecycle is presented in Figure 1.5-1. It portrays the use of this approach in cases where the overall systems architecture has been established and prototyping is being used to implement a preselected subset of the modules.

The "bottom line" on prototyping appears to be that it is a powerful means of defining and refining our knowledge about system requirements and improving communications with the client. However, it requires the utmost discipline in the project team to prevent runaway software development.

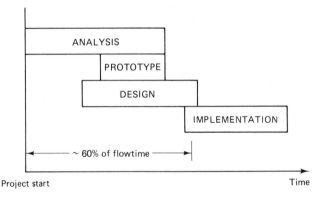

Figure 1.5-1:   Example of the Prototype Lifecycle Model [S1]

1.6 LIFECYCLES AND THE STRUCTURED METHODS

Structured Analysis and Structured Design have been used in conjunction with all the lifecycles and variations of them that we have discussed. The degree of success that different organizations have had in applying these concepts over the life of a software system depends on circumstances. The record is varied. Some have seen the methods as marvelously successful, while others felt that they inhibited progress. One reason for this situation is that the role of methods can be overemphasized; they can be seen—erroneously—as guarantors of success. Actually, success in software development is a direct result of a *system* of management and technical practices and procedures and *not* the result of a *secret*.

Having made these precautionary observations, let us take a look at the role that the structured methods play in the software development process. We will examine briefly each of the classic phases in the Software Lifecycle—its inputs, outputs, and its purpose. We will also identify the structured methods that play a key role in each phase.

1.6.1 The Analysis Phase

The goal of this phase is to build a problem model—that is, to create a description of just what is desired and what will eventually be built. In order to do this, the analysis phase requires a specific set of inputs. These may take the form of interviews, specifications as to level of performance, and random data. In addition, a considerable amount of information must be amassed regarding the client's policies: after all, it is the client's policies which will end up being implemented in the system.

The analysis phase goes through two major stages:

Physical model. This involves an examination of the current system through an analysis of its physical properties—for example, the number and types of forms and their content, use, and flow through the client's current system.

Logical model. This involves the abstraction or logicalization of the physical model into one which has had the mechanisms (the means by which tasks are accomplished) removed, together with many physical details. For example, in a physical model we may refer to a form as ''#123-pink copy,'' but in the logical phase this may become ''customer receipt.'' Note that we have isolated or identified the *function* or purpose of the form and removed its physical characteristics.

The desired output of this phase is a concise statement of *what* the system must do. The direction of this statement is *away* from how any of these tasks or functions will be accomplished and *toward* what is required. The analysis phase is also referred to as the *requirements* or *specification* phase—terminology that indicates the growing trend toward using the results of the analysis as the basis for design. Hence, a statement of requirements is a brief statement of the problem model (analysis). The analysis phase is depicted graphically in Figure 1.6.1-1.

Figure 1.6.1-1:   Graphic Model of the Analysis Phase

1.6.2 The Design Phase

The goal of this phase is the construction of a solution model. Note that this is only a *model* of the solution, not the solution itself. Obviously, the solution itself is the delivered system. Inputs to this phase include the output of the analysis phase, experience and system knowledge, and the method(s) to be applied in arriving at a solution model.

The design phase proceeds through two primary stages:

Logical Design. In this phase, often referred to as preliminary design, we create a design which will satisfy what was specified in the analysis phase. This design, however, does not include implementation considerations, constraints, programming language features, or other details of what the physical system will be like; it is for an abstract machine executing an abstract language. What good is it, then—since we will eventually have to implement it in the "real" or physical world? The answer lies in the concept of stepwise iterative refinement. That is, we make sure that we have a working concept that seems reasonable and meets a set of requirements *before* we attempt to implement it. This working concept is similar to the artist's conceptual drawing that is produced early in the development of a major construction project. The drawing is not exactly the way the building will eventually look; its dimensions will be slightly different, owing to cost and other considerations. The overall appearance, however, is close enough for purposes of decision making, refinement, and further investigation and physicalization. The logical design phase plays the same role in software development. An additional advantage to the use of logical designs is the fact that they are very easy to change as compared to physical ones or code.

Physical Design. This phase, also referred to as detailed design, produces the final blueprint for the system. Whereas the logical design phase may have resulted in the identification of a certain population of modules and corresponding control structure, the physical design phase takes them and applies constraints, details of language and hardware, and accepted software engineering practice. The number of modules that finally emerge from the physicalization process will, in general, not be the same as the number identified in the logical design phase. One reason is that some of the functions that were identified in the logical design phase may be part of the programming language selected or the operating system.

The information flow through the design phase is shown in Figure 1.6.2-1.

1.6.3 Implementation Phase

In some situations the implementation phase can be the simplest of the lot. In others—for example, if we choose to "Slam Dunk" the system—it can be a traumatic experience.

The goal in this phase is to implement the system according to the blueprint set down in the physical design phase. Granted, there may be nuances of which we were not aware in the physical design phase which cause us to want to change the blueprint, but that is to be expected. Something similar happens in the construction

Figure 1.6.2-1: Information Flow through the Design Phase

industry—but there, the construction people are required to get the blueprints reexamined and approved by local authorities *before* implementing the change. Similarly, the implementers of the software may wish to make changes. A configuration control approach enables the development team both to make such changes and to maintain the integrity of the software blueprint.

A graphic display of the information flow associated with the implementation phase is presented in Figure 1.6.3-1.

Figure 1.6.3-1:   Information Flow through the Implementation Phase

1.6.4 The Maintenance Phase

The maintenance phase is truly a miniature development cycle. A need is identified, it is analyzed, a modification is designed, and the change is implemented. Hence, it is a composite of the other three phases.

1.7 METHODS AND PHASES

The relationship between the structured methods and the phases just described is presented in Figure 1.7-1. The information flow among phases is shown in Figure 1.7-2. Note in Figure 1.7-2 that the information modeling activity is not singled out as a separate process. The reason is that its results and impact cut through both the analysis and the design activities. In a sense, those activities or phases are vertical, while information modeling is horizontal. A further symptom of this interaction is that the results of the process-oriented activities (e.g., the analysis and design of what will become code) will result in products which are highly likely to change, while the results of information modeling are likely to remain unchanged (or nearly so) over long periods of time.

Method	Phase
Structured Analysis	Analysis
Information Modeling	Analysis
Event Modeling	Analysis, design
Structured Design	Design
Logical Database Design	Design
Structured Programming	Implementation
Maintenance	All

Figure 1.7-1:   Relationship between Methods and Phases

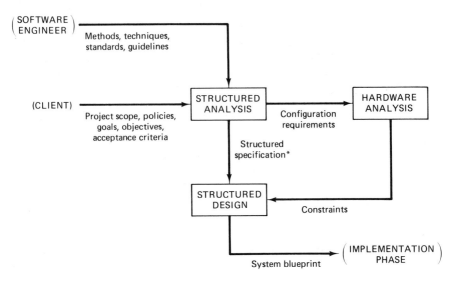

*Consisting of a Process Model, an Information Model, and an Event Model

Figure 1.7-2:   Information Flow Among the Phases [S1]

REFERENCES

1. P. Freeman, "Using the 'Right' Lifecycle Model," *Interface*, 1984.

2. D. Couger and R. A. Zawacki, "Key Factors for Motivating Computer Professionals," *System Analysis and Design: A Foundation for the 1980s*, edited by W. W. Cotterman et al., North Holland Publishing, New York, 1980, pp. 417–428.

3. G. Weinberg, *The Psychology of Computer Programming*. New York: Van Nostrand-Rheinhold, 1971.

S1. Extracted from the seminar, "Structured Analysis for Real-Time Systems," by Software Consultants International, Ltd., Kent, Washington, Copyright 1985, 1986. Reprinted by permission.

CHAPTER 2

A Systems View
of Software Development

From the very beginning, the computer industry has focused on the *results* of the development effort—the code. Despite a considerable amount of change in hardware and software technology, this initial view has been very difficult to modify. As an industry, we continue to develop newer, more powerful programming languages and computers but still fail to require that students learn to analyze and design *before* they write their first program. Even the methods that have evolved over the last twenty years reflect a continuation of the code-oriented view.

The earliest attempts to address concerns over software quality employed techniques generally referred to as "Structured Programming." Guidelines were established on the size of programs, the types and number of comments, indentation of code to highlight nesting levels, the use of the GOTO statement, and other techniques. All of these had positive effects on the code itself. However, they did not change the frequency with which quality code was produced, which solved a problem *other* than what the client needed solved. They also did not change the fact that relationships were created within the programs or systems of programs that made them difficult and expensive to maintain.

The architectural issues were addressed with the advent of a literal plethora of software design methods. Although all of these were touted as being "new," in fact they represented a set of design principles which were already accepted in various branches of engineering. For example, the decoupling of various functions is a well-known means of improving the reliability of a hardware system. Hence, when one component fails, the functions performed by most other components are not jeopardized. We "discovered" this principle in the seventies and claimed it as our own.

These "new" software design methods allowed their users to compose a software architecture having properties which the method's author(s) felt were important. Initially, there was no general agreement on just what properties were desirable and which were not. The situation has improved somewhat, but there still is no general agreement on the specifics of quality design. Despite the "advances," no strong link was established between the client's requirements and what was designed and, eventually, built.

The specification and/or requirements issues were addressed still later with the creation of a number of software systems analysis methods. Although the software field again claimed to have "discovered" such graphic schemes as the dataflow diagram, all of them had their origin in other branches of engineering. These methods did help to bridge the gap between the client's needs and what would be designed and built. But a new type of problem began to appear.

The definition and organization of the information to which the software would provide the user access began to be a serious problem in the last decade. Actually, the problem was always there, but only recently did the number and types of applications and queries evolve to the point where the oversights of the past began to take their toll. This resulted in the development of methods for defining information and the relationships that must be supported in the database.

Throughout its brief history, then, the software development field has been successful at identifying a specific problem and devising some approach to address it. Only recently have the elements required to create and support an integrated view of systems development been available. In this chapter, we will propose an integrated model for software development which incorporates all three aspects of software systems development. The three aspects we will be integrating are:

The process-oriented view ("process view"). This deals primarily with the coding and processing or transformations of data views.

The event-oriented view ("event view"). This deals with the real-time aspects of the software system.

The information-oriented view ("information view"). This deals with the

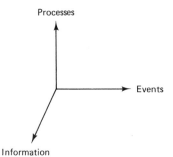

Figure 2.0-1:   The "Systems Space" View [S1]

definition of the information the software must process as well as the inherent relationships that must be supported by the database system.

These three aspects form a three-dimensional "systems space" in which each view is treated as a dimension (Figure 2.0-1). Each will be examined in more detail in the sections that follow.

2.1 THE PROCESS VIEW

This is the most common view of software and data processing systems. It is closely related to the original coding view, and most of its elements have evolved from code considerations. Examples of tools used to develop the process view include:

Flowchart and "structured programming" versions of it [1]
Input-process-output (IPO) diagrams [2]
Nassi-Shneiderman chart [3]

The basic notion behind this view is depicted in Figure 2.1-1. In it, the system is viewed strictly as a transformer or processor of information. This processor changes input information into output information in either of two ways: physical or logical. Each of these is discussed in more detail in the chapter dealing with dataflow diagrams.

Figure 2.1-1:   Graphic Depiction of the Process View

2.2 THE EVENT VIEW

Many of those who develop real-time systems find that today's software development methods do not provide what they need to describe such systems. Among the concerns of these software engineers are:

The population of states or modes of operations of the real-time system
The events that can trigger the system to "jump" from one state to another
The hierarchical relationships between states such that there are primary, secondary, etc., states or substates
The often mutually exclusive nature of some states while others may coexist (i.e., parallel states)

The use of graph or automata theory has helped somewhat, but the diverse nature of the population of software engineers makes it unlikely that these formal methods will be as widely accepted as they need to be. What is needed is a simple, graphical means of documenting the properties of real-time systems as described above.

More important, the event view must be an integral part of a systems-oriented view. That is, it must be complete and consistent with respect to the issues it addresses *and* consistent with the process view and the information view. An example of an event-oriented graphical approach is presented in Figure 2.2-1.

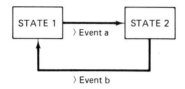

Figure 2.2-1:   Graphic Depiction of the Event View [S1]

2.3 THE INFORMATION VIEW

This is the last view to be considered from an industry standpoint, perhaps because its role in the software engineering process is not well understood. Its role is to *support* the development of databases as well as other aspects of systems development. It is *not* intended to be a database design method. The information view provides the software engineer with a means of capturing the facts about information that is so vital to the understanding and documenting of the characteristics of what may be loosely termed "data."

An example of the kind of discipline that is required in the information view, but not in the process view, is the definition and understanding of data. A simple term like "policy_holder" can be used in a program and database without any agreement among the insurance-company divisions using the results of the software as to what the term's definition is. This situation can go on indefinitely until two or more of the divisions are consolidated. Then the fact that each division has its own definition and its own set of programs and databases will cause problems. Even without a consolidation, there can be problems. For example, an attempt to provide *all* of the insurance and investment needs of the policy_holders may result in the discovery that there is a considerable amount of duplication, misinformation, and missing information in these multiple occurrences of policy_holder.

2.4 SUMMARY

All three views are needed to fully understand and document a software system. Software engineers have spent too long dealing with just the process view. It is the view that is the *most* likely to change, because it primarily results in code. Code reflects a response to competition, market pressure, the need for increased profits, and, sometimes, almost whimsical changes in policy. Although the process view does not represent much of an investment, it has enabled us to develop systems that work (after a fashion) but has actually prevented us from realizing the full potential of automation.

This message is brought home ever so clearly when we consider a real-time information system in which large volumes of data will be examined in unpredictable ways with certain processes being interruptable. The complexity of such a situation can quickly become overwhelming. This was brought home a few years ago on a satellite command and control system. Early versions of the system revealed that a serious timing problem existed. Nearly everyone agreed that it must be due to some slow-running, possibly poorly coded sections of assembly code. A firm was engaged to run a code analyzer on the system. The results indicated that the problem had nothing to do with the code per se but was due entirely to a complex data buffering scheme. Some people were so convinced that this could not be correct that they wanted the company to get its money back. Corporate attorneys advised that the remedial action(s) suggested by the analysis be taken. If the problem did not go away, then the company would have a strong case. The changes were made. The problem went away. Even with the advantages of hindsight, it is still not easy to see how such changes could produce such benefits—but they did.

In a form of "backlash" to the advent of methods and an engineering approach to software development, some people are seeking to find some other means to continue "slam dunking" code without really calling it that. The method often used is to "prototype" or get something up and running quickly and use that knowledge to expand and refine the capabilities of the system. The problem is, this is being done more and more in a "hobbyist" type of environment. That is, the concept of fixed price and fixed delivery schedule is being replaced with an experimental approach. While this may have some merit, the fact remains that development in this way lacks goals, definable controls, and the necessary feedback to enable an outside party to objectively determine what the status of the project is.

As we proceed with the rest of this text, we shall show that the viewpoint espoused here represents the first integrated means of bringing together what were formerly diverse views, employing a notation specifically designed to support this consolidation of concepts.

Perhaps the main point to keep in mind as you proceed through this text is that we will not address either real-time systems or database systems as separate and special types of systems. *All* software systems have properties requiring all three views. Certainly, some are degenerate cases, as is the case with batch systems, since they have only a single state (the minimum). Nevertheless, they do have a state. Thus, we will address the issues related to Structured Analysis and Structured Design as analysis and design issues—not as issues related to one special type of system or another. A practical reason is that it is getting harder and harder to make the distinction between one type of system and another. A more important reason is that to treat certain systems differently than others is to say that we need to have separate, independent analysis and design tools, each of which is special-purpose rather than generally applicable.

Throughout the remainder of this text we will refer to the relationships which exist among all of the tools of Structured Analysis and Structured Design as they have been extended to meet today's marketplace. These relationships were summarized in Figure P-1, which is presented again here as Figure 2.4-1.

TYPE OF MODEL	DEVELOPMENT PHASE		
	STRUCTURED ANALYSIS	STRUCTURED DESIGN	IMPLEMENTATION
PROCESS	Dataflow Diagrams Dictionary[1] Pseudocode	Structure Chart Dictionary[2] Pseudocode[4]	Code Dictionary[3]
INFORMATION	Entity-Relationship Diagram Dictionary[1]	Database Design Dictionary[2]	Implemented Database Dictionary[3]
EVENT	Event Model[5] Dictionary[1]	Event Model[6] Dictionary[2]	Queuing Model Dictionary[3]

[1] This dictionary includes data definitions: pseudocode for each process in the analysis model; descriptions, attributes, and content for the entities and relationships in the E-R diagrams; and descriptions of the events and states in the Event Model.

[2] This dictionary contains everything found in the analysis version of the dictionary plus definitions of all flags and pseudocode for all modules that were added as part of the design process or not otherwise present in the analysis.

[3] This dictionary has been updated to incorporate all information developed during the design activity which has been modified as a result of implementation considerations.

[4] The pseudocode referred to here incorporates any and all flags employed in the Structure Chart. All pseudocode is contained in the dictionary as well. Much of this may be identical to or based on the pseudocode that was developed during the analysis.

[5] Event Model refers to a simplified form of state transition diagram.

[6] This Event Model would be further refined and leveled based on design "discoveries."

Figure 2.4-1:   Relationship of Development Phases to Tools and Methods [S1]

REFERENCES

1. N. Chapin, "Flowcharting with the ANSI Standard: A Tutorial," *ACM Computing Surveys,* Vol. 2, No. 2 (June 1970), pp. 119–46.

2. "HIPO—Design Aid and Documentation Technique," IBM Corp., Manual No. GC20-1851, White Plains, N.Y.

3. I. Nassi and B. Shneiderman, "Flowchart Techniques for Structured Programming," *ACM SIGPLAN Notices,* Vol. 8, No. 8 (August 1973), pp. 12–26.

PART II

Structured Analysis

Structured Analysis has received quite a bit of attention in the software technical and managerial press. It represents a significant shift away from a procedural, sequential view of systems and toward an information-oriented one. But more than this, Structured Analysis emphasizes disciplined use of multiple means of expression in order to increase communication with a client. Its underlying concept is that the definition of a "new" system must be based upon an iterated and modified model of an existing system. This existing system may just be a set of policies which management wants carried out or it may actually be a physical system with forms, procedures, and, possibly, software. Hence, this new system will not simply "spring up" out of nowhere with no prior model to work from.

We will use the term "model" a great deal throughout this section and the rest of this book. It may sound abstract, but it isn't. It turns out that we use models in our daily lives to recognize people, places, and activities. Similarly, in Structured Analysis, we will emphasize the creation and refinement of various models in order to help us, as analysts and software engineers, to better understand what it is we are dealing with. These models also help us to identify characteristics about the existing system which are causing our client problems. These characteristics need to be modified while others need to be retained.

What we are attempting in all of this is to migrate in a disciplined manner from the way things are being done now to the way in which the client wants things to be done in the future. There will be a tendency to retain a good deal of what is currently being done because it is familiar. This is true even of many of the features that are causing the client problems. However, the techniques used in Structured Analysis make it relatively painless for the client to depart from what is toward what will be. Structured Analysis also provides an excellent starting point for the design phase, if we choose to automate any or all of the functions identified in analysis.

These last two points need further explanation. It is a common misconception that just because we apply Structured Analysis to some system, we must automate it. Quite the contrary: many manual and partially automated systems have been greatly improved through the increased understanding provided by Structured Analysis. This is possible because Structured Analysis provides us with a viewpoint that is different from the more accepted practice of describing systems by means of sequential models, such as flowcharts. Just as in the case of walkthroughs, an alternative viewpoint can be invaluable.

The second point, regarding design, is a definite money saver. The transition between specification or requirements definition and the design and implementation activities represents a significant problem. There is always the danger of losing a requirement, adding one unintentionally, or misinterpreting requirements. One value of having the requirements statement dictating the initial design architecture is its ensuring that, at least initially, the *structure* of the design will parallel that of the problem. Hence, we can meet at least one criterion of effective design-problem structure—that the problem structure be reflected in the solution structure.

In this section we will describe Structured Analysis—its basis, notation, and use. Topics which are not usually considered part of Structured Analysis, such as the use of supplementary methods and techniques as well as modifications to the original content of Structured Analysis, have been included. These additional techniques were necessitated by inadequacies of the original form of Structured Analysis.

The intent is to empower the reader to effectively address the analysis of just about any kind of system. This is why we have not chosen to single out "Structured Analysis of Real-Time Systems" as a topic. The philosophy here is that an effective analyst and software engineer should be able to accurately model *any* type of system. The particular type of system that you may be interested in may not be mentioned here, but enough types of systems and techniques will be discussed to enable you to contrive some combination of methods which will be effective.

This extension of the original content of Structured Analysis emerged from a systems-oriented view of this problem. From a systems standpoint, software (and other) systems display three types of characteristics:

Information. This aspect includes all properties related to dictionary, database definition and design, objects, entities, attributes, and relationships between entities.

Processes. This aspect includes the procedural properties of the system, including the definition of processes, the dataflows that provide them with the information they are transforming, and the pseudocode describing their operation.

Events. This aspect includes the description and documentation of the various states that many software systems may reside in, the occurrences that can cause them to change their mode of operation, and the compatibility of these states with each other.

These three aspects—Process, Event, and Information—constitute dimensions of a "space" which contains the software system. The remainder of this book will employ this view as a concise means of organizing the material and presenting information. Details of the tools and phases associated with this "space" are presented in Figure II-1.

Before proceeding, we need to address two other topics: change and information aspects.

The issue of change is crucial. Change is inevitable. Either we will accommodate it during the analysis phase or we will become its victims. Although it may not

TYPE OF MODEL	DEVELOPMENT PHASE		
	STRUCTURED ANALYSIS	STRUCTURED DESIGN	IMPLEMENTATION
PROCESS	Dataflow Diagrams Dictionary[1] Pseudocode	Structure Chart Dictionary[2] Pseudocode[4]	Code Dictionary[3]
INFORMATION	Entity-Relationship Diagram Dictionary[1]	Database Design Dictionary[2]	Implemented Database Dictionary[3]
EVENT	Event Model[5] Dictionary[1]	Event Model[6] Dictionary[2]	Queuing Model Dictionary[3]

[1] This dictionary includes data definitions: pseudocode for each process in the analysis model; descriptions, attributes, and content for the entities and relationships in the E–R diagrams; and descriptions of the events and states in the Event Model.

[2] This dictionary contains everything found in the analysis version of the dictionary plus definitions of all flags and pseudocode for all modules that were added as part of the design process or not otherwise present in the analysis.

[3] This dictionary has been updated to incorporate all information developed during the design activity which has been modified as a result of implementation considerations.

[4] The pseudocode referred to here incorporates any and all flags employed in the Structure Chart. All pseudocode is contained in the dictionary as well. Much of this may be identical to or based on the pseudocode that was developed during the analysis.

[5] Event Model refers to a simplified form of state transition diagram.

[6] This Event Model would be further refined and leveled based on design "discoveries."

Figure II-1: Relationship of Phases to Tools and Model Types (Extracted from the seminar, "Structured Analysis for Real-Time Systems," by Software Consultants International, Ltd.; Kent, Washington; Copyright 1988. Reprinted by permission.)

always be evident, the remaining chapters in this part and Part III assume that there will be nearly continual change present. This change will be such that almost nothing is stable or sacred within the system. This is why the emphasis in this section is on the gathering and documentation of information about information.

Information is, in general, the most stable aspect about systems. Hence, some degree of stability can be achieved by addressing information issues in a primary rather than a secondary manner. This can, however, create some problems. One of them is that currently, and for the foreseeable future, we will be dealing with project managers and clients who are largely process-oriented. That is, they view a system development effort as being directed primarily at the production of a product— code. They do not see the code as merely a logical extension of information definition and policy into the *process* dimension. Hence, if we tell such a person that we are going to spend the first half to two-thirds of the analysis phase defining information, forming an information model, and documenting policies about information, such a person may issue an edict to "Slam Dunk" the system and cut the science! In fact, in some projects that is *exactly* what has happened.

The terrain being what it is, we should observe the motto attributed to the Swiss army: "When the map and the terrain disagree, trust the terrain." We can conceive of a day when the information side of analysis will dominate analysis activity, but that is the map. Until it does, we must face the realities presented by the terrain. Some consideration must be given to process characteristics early on in analysis. Thus, we will examine some processing aspects and use them as a means of deriving information ones. This seems the "right way" to most managers at this time. Hopefully, a future revision of this text will be necessary when this view gives way to one which addresses the information aspects first.

Certainly, both aspects are necessary. However, the quality of information-modeling results can be degraded somewhat by considering the process aspects first. An alternative is to do them independently, then bring them together at the end for consistency checking. Anyone who has really tried this strategy realizes that it is a lot more work in practice than textbook authors are willing to admit in writing— enough work to warrant avoiding the strategy if at all possible.

In this part of the book, our primary theme is that all three aspects of systems analysis are necessary. We did not choose to separate out information modeling into its own chapters, because essentially it cuts through all aspects of analysis and design. Addressing it as a separate topic would be the same sort of error that has plagued Structured Analysis and Structured Design for years—considering information issues as a separate topic.

CHAPTER 3

The Lifecycle Dictionary

The term *data dictionary* has come to have a dual meaning. Among vendors of database management systems (DBMS), the term refers to a catalog of data items and their properties that the DBMS must manipulate. This catalog is oriented toward the support of the DBMS. As a result, the emphasis in such dictionaries is on the physical characteristics of the data (e.g., fields, data type, size of fields). Although this is important implementation information, it does not really address the needs of the analysis phase. Those needs include the definition of what data names mean, what other data items they are composed of, what the company policy is regarding that data, where each data item comes from, and where it goes to. It is often the case during the analysis phase that the physical properties of the data have not been established. This is especially true when a new system is being defined. Hence, a DBMS data dictionary actually asks the "wrong" questions during the analysis phase.

Those who have come to know the term data dictionary from other works on Structured Analysis probably feel that that concept asks the "right" questions. These "right" questions are related to the composition of the data, uniqueness and consistency of names, and definitions of terms. This is more appropriate to the analysis activity—particularly when it comes to the definition of terms.

In the case of the Structured Methods, data dictionaries are employed as a means of defining terms and enhancing communications. This is in keeping with the industrywide shift in emphasis from sequential logic and code to data and information. Our view of systems now reflects a perception of them as processors of data and implementers of policy. In attempting to gain an understanding of the processor, we must not overlook the data being processed. Often, systems analysts have tried to employ the Structured Analysis concept and not used the data dictionary. One reason (which we will expand upon later) is that the data dictionary can include a very large amount of information to document and maintain. When documentation and maintenance are done manually, things quickly get out of hand. Even so, abandoning the use of the data dictionary only aggravates the very communication problems we are attempting to overcome. That is, without understanding the data, we can hardly hope to understand the system which processes it, the partitioning of those processes, and the system's relationship to the client. This understanding requires accurate definitions which are open to criticism and refinement.

Without definitions, we can find ourselves in a "Tower of Babel" situation in which we think we know what is being said but we do not know what was meant. This can occur even in situations where we are familiar with the type of business our client is involved in. This is illustrated by the experience of a flight instructor. He had a military background and had also instructed literally hundreds of private pilots of multiengined aircraft. One day, while training a pilot how to land a B-17 which had been converted for use in fighting forest fires, he was aware that they were coming in low and slow. Since he believed that he should give the trainee as much time as he could in order to see if he could detect and correct the problem on his own, he waited patiently. Unfortunately the student pilot did not notice the problem, and things got considerably worse. Finally, the instructor could wait no longer and shouted, "Takeoff power!" In the military, the instructor had become familiar with this term as meaning to apply full power to the engines, but the student did not have the same background. Instead, he interpreted the admonition to mean just the opposite (i.e., take off power), so he cut the engines still further back. Fortunately, both parties survived the ensuing crash, although not much of the plane did. The next time you wish to cut corners by "skipping" the dictionary, recall the term "takeoff power" and see if that doesn't change your mind.

We have already discussed the fact that the results of the analysis activity will have a strong, if not dominating, influence on the quality of the design. If that design is to emulate the kinds of structured relationships present in the system, it should be driven by a stable element in that system. Data is a very stable element in a system. For example, the way in which telephone calls are processed has changed dramatically over the last 20 years, but the data processed (i.e., area code, exchange number, and subscriber number) has not. If we are going to base a design on a process-oriented model, then we need a stabilizing influence. The data dictionary is a necessary part of defining and utilizing that influence.

In this text we shall employ the term "Lifecycle Dictionary." It will refer to the collection of information that enters, leaves, and is transformed by the system of interest, the policies surrounding that information, and descriptions of *all* other objects or events of interest with respect to the system. This includes data composition, policies about data, event and state descriptions, entity descriptions, relationship descriptions, pseudocode and/or structured English, policies associated with these different types of information, and any other related information deemed useful. If we were to take this concept to its logical limits, the Lifecycle Dictionary would also include the actual code, test plan, test cases, test data results, results of walkthroughs, and just about anything else associated with the project. However, for the purposes of this text, and in introducing this somewhat radical concept, we will emphasize the first list of information. A Lifecycle Dictionary, then, as referred to here, is a repository of information about information. Without the application of a considerable amount of discipline, in the form of standards, guidelines, and procedures, this repository can become a "garbage can."

Throughout this book, we will use the terms "dictionary" and "Lifecycle Dictionary" interchangeably, because the "data"-oriented view of the dictionary is far too restrictive. There exists a considerable amount of information regarding the models and our project which just does not conveniently belong in the "data dictionary." For example, in a payroll system, there may be a definition of what a pay code is in the data dictionary. Where would we put the policy related to who authorized it, who can access it or change it? Also, where might we conveniently store and control changes in structured English or pseudocode? The dictionary is a cost-effective way to do this. Over time, it can evolve into an invaluable time- and money-saving entity used by one new maintenance effort after another. To make its role clearer, think of it as a repository for *all* information related to the project. This includes management and scheduling information as well as the more technically oriented information. Certainly, only a subset of the contents of the Lifecycle Dictionary is reusable—but what a subset! Literally *all* of the definitions of data and policies related to it and its attributes will be reusable. It can evolve into a complete model of *all* information and policies related to the enterprise. Thus, within the span of a very few projects, the possibility exists that no new information will have to be entered into the Lifecycle Dictionary with respect to data, policies, and attributes. More on this topic later when we discuss information modeling.

The primary reason for discussing this topic so early is to recognize its importance to the successful use of the structured methods—*any* structured method. If we do not put communication through definition, documentation, and concurrence at the top of our priority list, then we are in for serious, long-term problems.

3.1 THE ROLE OF THE LIFECYCLE DICTIONARY WITHIN STRUCTURED ANALYSIS

In reviewing the results of a Structured Analysis effort, remember this: if no Lifecycle Dictionary has been done, then all we are reviewing is a bunch of pretty pictures—*not* analysis. The use of the Lifecycle Dictionary is fundamental to the use of Structured Analysis and, incidentally, to the use of Structured Design. It is the only link which solidly ties the analysis work to the design. Without it, the design's pedigree is not only questionable but almost nonexistent.

The importance of the Lifecycle Dictionary to Structured Analysis will be clearer if we examine the nature of the businesses we are analyzing. Most business policy is the result of tactical responses to changes in government regulations, competition, or corporate goals. The implementation of these policies involves language—that is, words, names, titles, labels. As changes occur and the corporation grows, a certain amount of corporate regionalism begins to grow as well. From a Lifecycle Dictionary standpoint, this means that local meanings for various terms start to appear. Since Lifecycle Dictionaries have been around only for a short time and businesses for a much longer time, Lifecycle Dictionary situations which appear illogical and highly improbable abound. Some examples:

An insurance company's various divisions cannot come to an agreement on a definition for the term "policyholder."

A telephone company's departments cannot agree on a definition for the term "subscriber."

A government agency's departments cannot agree on a definition of "travel voucher."

A manufacturer's production divisions cannot agree on the meaning of the term "manufacturing plan."

There appears to be a natural law of effective analysis, which may be stated as follows:

Whenever a system is analyzed, some previously known aspect of that system will be discovered and manifest itself (at least in part) in terms of data definition.

Another natural law is related to the manner in which we have been developing software systems over the last thirty or more years. It is that much of the confusion and expense that organizations experience in changing current systems to meet an ever-changing environment is due to a misemphasis at the time the system was built. Emphasis was on the *product* of the system rather than on the information handled by that system. Recently, an insurance company experienced this situation. Every time they wished to change an existing system to accommodate some new need, the change turned into a major project. These projects often required several months and several people to accomplish. The company's history of systems development was one of building a system to meet a need. Hence, each new type of service and each new type of coverage often resulted in the construction of a new system. Later, as the company and governing laws migrated into new service areas (e.g., investment management, stocks, trusts, IRAs), these existing systems had to be modified and, in many cases, had to share or jointly utilize data. This aggravated the situation created by independent system development.

Another natural law may be in order here:

Long-term maintainability of software systems is possible *only* if the focus of attention during development is on the information going into and out of the system, the relationships inherent in that information, and the relationship of that information to the rest of the enterprise (i.e., business, agency, division).

The system aspects "discovered" upon analysis will remain undocumented and unresolved unless the Lifecycle Dictionary is employed. Certainly, resolution of problems like those cited above may take some time, may involve the top management levels of the company, or may just remain unresolved by management. The point is: if it is not documented, it does not exist!

An additional point widely misunderstood about the Analysis Lifecycle Dictionary is exactly what it should and should not contain. Its primary purpose is to document the relationships between/among data elements, their composition, their names. Its primary purpose is *not* the documentation of physical format, ranges of values, method of computation, or other ancillary information; these are in the domain of the Design Lifecycle Dictionary and will be discussed later. What one author [1] recommends be included is summarized in Figure 3.1-1.

INCLUDE

NAME	(As known by customer)
COMPOSITION	(e.g. "Client name is composed of an optional TITLE, FIRST INITIAL, AND FAMILY NAME)
APPROPRIATE	("Due to be changed 1 Jan 84 – TITLE
COMMENTS	will be DELETED") three references are required
MAX/MIN	(e.g. "At least one and as many as
NUMBER of	three references are required)
ITERATIONS	
POLICY	(e.g. "only the Chief Executive Officer may access this information")

EXCLUDE

    PHYSICAL FORMAT*

    RANGE OF VALUES*

    METHOD OF COMPUTATION (i.e. no formulas)*

*We would include this information only if we were replacing an existing system. It would also be clearly flagged as being part of the current system and not, necessarily, the way the new system will be set up.

Figure 3.1-1:   One Approach to S/A *Data* Dictionary

A counterargument to this guideline relates to the analysis of existing systems which are, at least partially, automated or will interface with an existing system which is automated. A lot of information about these systems has not been written down. When this analysis effort comes along, we have, for the first time, the opportunity to "nail down" many "loose ends" associated with this system. Why not do it?

Therefore, there is something of an ambivalence about the content of the Lifecycle Dictionary. On the one hand, there is a danger of its becoming a "garbage can" of data about data. On the other, there is a danger of losing valuable data about data because we are not supposed to include it right now.

Let us look at the ways in which others have resolved this issue in actual practice:

Restrict the Analysis Lifecycle Dictionary No Matter What.   This approach follows the admonition of [1] to the letter. Only compositional information about data is included. The problems many have had with adhering to this position is that the data dictionary is not really respected as a project resource. Its content is limited and its value is perceived as highly limited. One side effect is failure to capture *policies* about data which would not otherwise be documented. For example, in a payroll system, this restricted view of data definition leaves us no effective way to document the fact that certain types of information (e.g., pay rate) may be examined only by certain persons and paycode may be changed only by certain employees.

Modify the recommended approach.   Many devotees of Structured Analysis have experienced problems with the first approach. As a workaround, they have found it useful to create an additional section to their data dictionary document, often referred to as a "hip pocket" or auxiliary data dictionary. This is a collection of all necessary additional information that would be useful during the software design phase.

Use a new concept.   This involves reexamining the entire analysis and design activity with an eye to a systems approach to documenting project-related information. We shall expand on this concept later, but it is a simple one. This view holds that the Lifecycle Dictionary is the best hope we have of creating a true project repository which is controllable and complete. The operating rule here is that *anything* related to the project (policies regarding data, formulae, value ranges, physical format) is included in this body of information. Two things must accompany the use of a Lifecycle Dictionary of this type:

1. Strict discipline regarding the inclusion of information, such that data regarding physical format, for example, will be included only if it actually exists or is required. That is, we are not speculating and prematurely implementing the system but are documenting a fact regarding it and/or a system it must interface with.

2. Automated support for the approach. It is difficult enough to use even the most restrictive approach to a Lifecycle Dictionary when manual methods are employed. When we expand the data volume that we have to deal with, the problems become overwhelming. Automation helps both the feasibility and the control problems.

3.2 NOTATION

The notation employed by the Structured Analysis approach is not necessarily unique. It is simple to learn and easy to implement on a line printer or word processing system. This suggested notation is capable of addressing the basic kinds of relationships which exist between/among data items and elements. These are:

> Concatenation
> Iteration
> Selection

In addition, other types of information enhance our ability to communicate definitions. These are:

> Comments
> Values
> Option(s)
> Composition

The symbols associated with each type of relationship are presented in Figure 3.2-1. The use of these concepts and their respective symbols is described in the remainder of this section.

NAME	SYMBOL	MEANING
composition	=	"is composed or consists of"
concatenation	+	"and"
iteration	{ }	"multiple occurrences of"
selection	[	]	"choice1 or choice2"
option	( )	"may or may not be present"
discrete value	" "	"the value of this variable"
comment	* *	"additional information"

Figure 3.2-1:   Structured Analysis Lifecycle Dictionary Notation

3.2.1 Concatenation

This is the most basic relationship represented in the Lifecycle Dictionary. It identifies what data elements or groups of data elements comprise a given entity. For example, instead of using text to express

> Customer mailing address is composed of number, street, city, state, and zip code

we could express it, using the shorthand notation previously described, as

```
CUSTOMER-MAILING-ADDRESS = NUMBER + STREET + CITY + STATE + ZIP_CODE
```

Note that the symbol + does not have the same connotation it has in mathematics. We are not computing customer address through a series of additions but using + to act as a delimiter to show composition by distinct elements.

Our example is somewhat restrictive. As we look at other notational symbols, we will accommodate other types of addresses.

3.2.2 Iteration

Often, several sets of elements may be allowed—lists, files, members of a group addressed by a single name. Going back to our previous example, if we allow customers to have more than one address, then

```
MAILING-ADDRESS = {NUMBER + STREET + (UNIT_NUMBER)
                   + CITY + STATE + ZIP_CODE}
```

Note that the symbol { } just indicates that several occurrences or values are possible. It puts no limits on these possibilities. The default values for iteration are zero for the lower limit and undefined (infinity) for the upper limit. Stated another way, what the above definition says is

> MAILING ADDRESS IS COMPOSED OF AS FEW AS ZERO AND AS MANY AS AN UNDEFINED NUMBER OF OCCURRENCES EACH CONSISTING OF NUMBER, STREET, OPTIONAL UNIT NUMBER, CITY, STATE, AND ZIP CODE

There are two common ways to notate the limits indicated above. Approach 1 is to present the lower and upper limits, when known, before and after the iteration symbols, respectively. Approach 2 is to place the upper and lower symbols above and below each other, respectively, in front of the first iteration symbol. These schemes are demonstrated in Figure 3.2.2-1.

Approach 1:

$$\text{MAILING-ADDRESS} = {}^{\text{upper limit}}_{\text{lower limit}} \left\{ \text{Street} + \text{City} + \text{State} + \text{Zip Code} \right\}$$

Approach 2 (the recommended approach):

$$\text{MAILING-ADDRESS} = \text{lower limit} \left\{ \text{Street} + \text{City} + \text{State} + \text{Zip Code} \right\} \text{ upper limit}$$

Figure 3.2.2-1:   Notation Used to Indicate Limits of Iteration

Although both approaches accomplish the same thing, many analysts prefer to use Approach 2. Experience has shown that whether the Lifecycle Dictionary is manual or automated, the linearity of this approach reduces confusion, and it seems to be more natural to more people to write.

3.2.3 Example of Part of the Lifecycle Dictionary Notation

The example we will use is based on the analysis of a service which shuttles people to and from the local airport [3]. The interviews revealed that there were several different kinds of clients, methods of payment, and factors involved in the creation, modification, and deletion of reservations. The resulting initial data dictionary entries are presented in Figure 3.2.3-1. Note that some policy has already been embedded in it. The exact format of this information is up to the user.

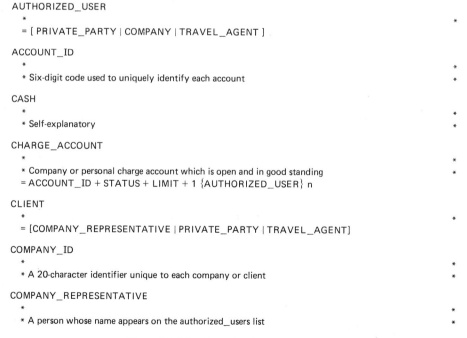

DATA DICTIONARY FOR THE BUS SERVICE

The dictionary entries presented below are provided as an example of a starting point for an analysis dictionary and are not meant to represent what may be possible with additional knowledge and insight into the business.

AUTHORIZED_USER
 *   *
 = [PRIVATE_PARTY | COMPANY | TRAVEL_AGENT]

ACCOUNT_ID
 *   *
 * Six-digit code used to uniquely identify each account   *

CASH
 *   *
 * Self-explanatory   *

CHARGE_ACCOUNT
 *   *
 * Company or personal charge account which is open and in good standing   *
 = ACCOUNT_ID + STATUS + LIMIT + 1 {AUTHORIZED_USER} n

CLIENT
 *   *
 = [COMPANY_REPRESENTATIVE | PRIVATE_PARTY | TRAVEL_AGENT]

COMPANY_ID
 *   *
 * A 20-character identifier unique to each company or client   *

COMPANY_REPRESENTATIVE
 *   *
 * A person whose name appears on the authorized_users list   *

Figure 3.2.3-1:   Example of *Data* Dictionary Use

CREDIT_CARD

  *   *
  * Only VISA and MASTERCARD   *

DATE_OF_RETURN

  *   *
  = MONTH + DAY + YEAR

DAY

  * Will be expressed in standard format — that is, dd where 01 ≤ dd ≤ 31

DEPARTURE

  *   *
  = DATE_OF_DEPARTURE + TIME_OF_DEPARTURE + PLACE_OF_DEPARTURE + FLIGHT_
    NO. + DESTINATION + DESIRED_TIME_OF_ARRIVAL

DESIRED_TIME_OF_ARRIVAL

  *   *
  * This is the time that the client wishes to arrive at the airport.   *
  * It is in the standard format: hhmm (hours, minutes) in Navy (24-hour-clock) time.

LIMIT

  *   *
  * The dollar limit that a given account has a credit limit. No reservations will be accepted if the current *
  * amount owed is equal to or greater than the LIMIT   *

METHOD_OF_PAYMENT

  *   *
  = [CASH | CREDIT_CARD | CHARGE_ACCOUNT]

MONTH

  * Will be expressed in standard format — that is, mm where 01 ≤ mm ≤ 12   *

PLACE_RETURNED_TO

  * This is the location that the client is to be returned to after being picked up   *

REQUEST

  *   *
  * A request for service made by a CLIENT directly or by a designee (e.g., a travel agent). The request can *
  * be for pickup from and or delivery to the airport.   *
  *   *
  = 1 [DEPARTURE | RETURN | DEPARTURE + RETURN] COMPANY_ID + METHOD_OF_PMT *
       + NAME_OF_TRAVELLER + DAY_PHONE_OF_TRAVELLER n

RETURN

  *   *
  = DATE_OF_RETURN + TIME_OF_RETURN + PLACE_RETURNED_TO + DESIRED_   *
    PICKUP_TIME + DESTINATION + DESIRED_ARRIVAL_TIME_AT_DESTINATION

STATUS

  *   *
  * Current classification of client's account   *
  *   *
  = ["OK" | 'CLOSED' | "OVERDUE"]

YEAR

  * Will be expressed in standard format — that is, yy where 00 ≤ yy ≤ 99   *

Figure 3.2.3-1:   continued

3.3 SOME SPECIAL PROBLEMS

Large and small companies can operate reasonably well, even though many misunderstandings exist about what is meant by or contained within a certain data name. These misunderstandings have hidden, negative effects on the company's productivity. The lack of precise interaction between parts of the company masks these effects. Not until an analysis is performed to support some degree of automation do these problems surface, requiring immediate attention.

For example, an insurance company discovered almost by accident that the meaning and use of the term ''salesman's commission'' varied dramatically between the company's various echelons.

Policy related to the computation and adjustment of commissions was not uniform throughout the company. As a result, commissions were often overpaid, underpaid, and paid multiple times. Much of this confusion stemmed from the fact that no definition of what was meant by commission was established and maintained. Each person in the company was free to interpret it as he or she saw fit. Also, company policy regarding the computation and/or award of commissions was not documented to a practical level. Many situations arose which were not covered. This let hearsay take control of this important aspect of the relationship between the company and salespersons. Such policy issues are discussed in more detail in Chapter 6.

The most common of the definition problems are aliasing and imposters, which are discussed below.

3.3.1 Aliases

Aliases (i.e., the same definition given to several items with different names) exist to some degree in every organization, whether they have used Structured Analysis or not. Just applying an analytic technique of this type does not mean that the political, cultural, and economic factors that produced the aliases will disappear. Sometimes they actually grow stronger and more defensive. Again, an alias exists when two or more data items have the same definition but have different names. Where do these names come from? Usually they arise because of a lack of central planning and policymaking. Independent organizations compose their own terms for certain items. Although this does cause some communication or interface problems between organizations, the company may operate in an apparently efficient manner until automation is considered. At that point, problems become apparent.

For example, one manufacturing firm wanted to automate its machined-parts tracking system. Occasionally a part that was made in its machine shop was defective in a way that could not cost-effectively be detected by quality control. It often happened that such problem parts could not be reworked on the production line. Hence, an expensive product could have its delivery date missed while the production line waited for another part of the right type. Such delays had serious consequences for both the manufacturer and the client. The proposed automated system would permit the identification of the status of the most complete part and enable the production supervisor to put a top priority on the completion of it. Aliasing almost sabotaged this effort. Here is how.

Each machined part began as a piece of stock (e.g., a block of aluminum) with an attached plastic bag. The bag contained a piece of paper that described the process by which the unmachined stock would be turned into the desired part. The name given to this piece of paper depended on what department you spoke with. For purposes of discussion, we will mention only three of them: manufacturing plan, shop work plan, and production specification. All had identically the same definition. Since the groups could not obtain agreement on a common name, they had an aliasing situation. Each group within the organization had adopted a name which worked well for them. Since these names did not have to be entered into a central system prior to the automation, there was no apparent problem. But once a common set of definitions were required, each group became defensive about their own respective name and refused to adopt the name used by any other group.

The best way to cope with aliasing is to eliminate it. For example, obtain concurrence on a common name. But what can we do if no concurrence is forthcoming? One thing we cannot do is to ignore it. If we do, then it is possible that one of the terms will have its definition changed while the others remain the same. Then a synchronization problem occurs. Changing all of them is also unacceptable, owing to increased labor. An effective way to mask this "disease" without curing it is to define one of the aliases and have the others refer back to it.

For example, let us take another look at our manufacturing situation, where three different names were used to define the same set of data elements. What we need to do is link these elements in a way that will ensure that when one changes, they will all change. Here is how we can do that. Merely define one of the names, and use comments to tell the reader what the aliases are. Then, list each alias but do

```
PRIMARY_NAME      = * AKA SECONDARY NAME   *
                    * AKA TERTIARY NAME     *
                    actual definition

          .
          .
          .

SECONDARY_NAME  = * SEE PRIMARY NAME        *
                  ( no definition here)

          .
          .
          .

TERTIARY_NAME   = * SEE PRIMARY NAME        *
                  ( no definition here)
```

Figure 3.3.1-1:   Using Notation to Control Aliasing

not define it. Merely use a comment to refer back to the primary name (Figure 3.3.1-1). The alias chosen for the role can be picked arbitrarily, be the most popular, etc. If any individual alias needs to be changed to become truly unique, just replace the comment for that data item with its (now unique) definition and remove the comment in the primary data item that refers to this (now former) alias.

This approach for dealing with aliases has the advantage that if a change is to be made to the primary data item, the analyst knows immediately and explicitly what other data item(s) will be effected. At that point consultation with the other analysts involved may be necessary in order to determine what action should be taken. This approach has several benefits. One is that the definitions for all aliases are synchronized. This is not very likely to happen if each has its own definition. Another is that if the definition of any of them is to change but the others are not to change, we can eliminate it from the a.k.a. Conversely, if the primary definition is to change, the analyst is made aware of the ramifications of the change (i.e., which other data items will be affected). The main thing to keep in mind about aliasing is that if we cannot eliminate it, we must control it.

Let us use our earlier manufacturing example. For the sake of discussion we will restrict ourselves to just three of the many names that were involved. Also, we will not detail the definitions themselves. The primary name that we will use is MANUFACTURING_PLAN. The other two names will be SHOP_WORK_ORDER and PRODUCTION_PLAN. The application of this concept to these names (and definitions) would result in the entries shown in Figure 3.3.1-2. Note that any change to MANUFACTURING_PLAN will affect the other two entries. However, the analyst making such a change would be forwarned about its potential impact. Conversely, anyone wishing to use either of the alternate definitions is pointed back to the primary definition. If it becomes necessary to change one of the

```
MANUFACTURING_PLAN  =  *AKA PRODUCTION_PLAN                    *
                       *AKA SHOP_WORK_ORDER                    *
                       (   actual definition of composition, etc.      )


                    .
                    .
                    .

PRODUCTION_PLAN     =  *SEE MANUFACTURING_PLAN                 *
                    .
                    .
                    .

SHOP_WORK_ORDER     =  *SEE MANUFACTURING_PLAN                 *
                    .
                    .
                    .
```

Figure 3.3.1-2:   Applying the Aliasing Control Notation

other definitions *only,* the pointers from the primary to it and from it to the primary are removed. Under such a condition, we have one less alias to concern ourselves with.

3.3.2 Imposters

Imposters (i.e., items having the same name but multiple or different definitions) are just the reverse of aliases, but they present us with a markedly different set of problems. The biggest one is that the names of these items have to be changed. Using what amounts to subscripting is probably the easiest means to an end.

3.3.3 Hybrid Data

Hybridizing of data seems to be one of the favorite problems that occurs in commercial applications. Hybrid data occurs when information has a "hidden" or mixed use. For example, a savings bank employed an account numbering system which incorporated the branch number in the first two positions:

```
BBNNNNNN

B = indicates branch number

N = indicates saver number
```

The problem occurred when the bank opened, or planned to open, its 100th branch. As a result, all accounts in the system had to be renumbered. This not only represented a large revenue loss for the bank, but also created several logistic problems as well. All of this stemmed from ignoring a simple rule about naming and dictionaries:

Use a name which tells us what the information is, not one that is used as a placeholder for different types of information

Actually, hybrid data results from ignoring the basic rule about systems analysis and development:

PLAN ON CHANGE!

3.3.4 Real-Time and Concurrent Processing Signaling

Many systems will require the use of one type of interrupt or another to cause the system to change its mode of operation. This results in some processes being turned "off" while others are turned "on." In order to make sure that we understand just what is involved and is intended, we need to document the interrupts that can occur and give them a name. Also, where we know just which types of processing should be ''on,'' this should be so indicated in the data dictionary entry. An interrupt might be documented using the following as an example:

```
SYSTEM_TEST   =  TYPE:              INTERRUPT
                 OCCURRENCE:        MAY BE REQUESTED AT ANY TIME BY
                                    THE OPERATOR
                 INTERRUPTABLE:     NOT UNTIL TEST IS COMPLETED
                                    SUCCESSFULLY OR UNSUCCESSFULLY
```

Other forms and approaches are possible. These will be discussed in more detail in the chapters on Structured Design.

3.3.5 Orphaned Data

In most cases, data is an orphan. That is, no one in the organization is responsible or accountable for its initial definition, future modification, and coordination of its use through subsequent projects. The concept of the data administration function only partially addresses this need. At the corporate level, there is an incredible amount of data to keep track of. But this corporate-level problem starts at a local level.

To help bring the problem under control locally, the author of the definition, date of last change, and other ''pedigree'' information should be recorded. In this way, those who need to have the definition altered know whom to see, the responsibility for each data item is established and maintained, and the continuity of responsibility can be handed to another person when the author leaves the organization.

Using our SYSTEM_TEST definition as an example, we can prevent its being orphaned by using the following definition:

```
SYSTEM_TEST   =  AUTHOR:

                 CREATION_DATE:

                 DATE_OF_LAST_MODIFICATION:

                 USED_BY:

                 TYPE:              INTERRUPT
                 OCCURRENCE:        MAY BE REQUESTED AT ANY TIME BY THE
                                    OPERATOR
                 INTERRUPTABLE:     NOT UNTIL TEST IS COMPLETED
                                    (SUCCESSFULLY OR UNSUCCESSFULLY)
```

3.4 LIFECYCLE DICTIONARY STANDARDS

We will describe two types of standards and their exceptions: conceptual and notational. The conceptual standards are directed at establishing the basis for the Lifecycle Dictionary while the notational ones relate experience and practice.

3.4.1 Conceptual Standards for the Lifecycle Dictionary

A set of conceptual standards that forms a basis for the dictionary consistent with the methods presented here are listed below.

The Lifecycle Dictionary contains definitions of data items, documentation of policy, and other project-related information.

1. All data and other information that comprise the analysis, design, and code must be defined in the Lifecycle Dictionary.

2. All data and other information that are contained in the Lifecycle Dictionary must be used in some part of the system.

3. The information in the Lifecycle Dictionary should be internally consistent, complete, nonredundant, and correct.

4. The Lifecycle Dictionary is intended to be the primary "Project Information Resource."

5. The Lifecycle Dictionary may contain comments regarding data that the software engineer deems necessary to understanding and using the information contained in it.

6. The names used for data items within the Lifecycle Dictionary and the names used on the dataflow diagrams must be identical.

7. The names used for data items within pseudocode must be identical to those used on the dataflow diagrams and the Lifecycle Dictionary.

8. The author of each definition should be identified in the dictionary.

9. The author of a given definition must be notified of and approve changes to it.

3.4.2 Notational Standards for the Lifecycle Dictionary

Composition of data items is described by utilizing a specific set of symbols. Each symbol connotes a type of compositional or relational function.

The symbols for use within the Lifecycle Dictionary are listed below, together with their meaning.

SYMBOL	MEANING
+	Concatenation
{ }	Iteration
[ ]	Selection
" "	Literal values
* *	Comments
( )	Options
=	Composition

A common set of name elements or abbreviations should be used throughout the dataflow diagrams, Lifecycle Dictionary, pseudocode, and entity-relationship diagrams.

3.5 CONTENTS OF THE LIFECYCLE DICTIONARY

Since the Lifecycle Dictionary amounts to a repository for *all* project-related information, there is little that should not be recorded in it. However, there is a minimum list of information which it should contain:

Pedigree Information. This includes author, creation date, validation procedure and date, date of last change, project(s) and system(s) used on, and requirement(s) on this project that this entry is related to.

Dataflow definitions. These describe the pipelines which transport information through the dataflow diagram.

Pseudocode descriptions. These describe the policies which are carried out by the processes in dataflow diagrams and the modules that comprise structure charts.

Event descriptions. These document the occurrences or circumstances that cause the system to change its mode of behavior.

State Descriptions. These document the processes which will be active when a particular event has occurred.

Datastore descriptions. These document the contents of the temporary holders of data which are an inherent part of the dataflow diagram.

Entity descriptions. These describe the groups of objects which are part of the Information Model.

Relationship definitions. These describe the nature of the relationships which the entities that are part of the Information Model share.

In all cases, policies regarding information should be included where appropriate. For example, if the definition of a payroll record includes the employee's pay code, and the definition of target description includes the target's latitude and longitude, who may access and examine these? Who may alter them? The narrow view of dictionary (namely, data dictionary) would have "policies" put into pseudocode. For some policies, however, it will not be obvious how they should be described in pseudocode. Many of these are procedural in nature. Hence, the safe and effective way of dealing with this situation is to document policy in a convenient and appropriate way. The broader view of the dictionary makes this a viable option.

3.6 AUTOMATING THE LIFECYCLE DICTIONARY

The Lifecycle Dictionary is the one aspect of Structured Analysis which is most often not used. The amount of change that occurs during its development, combined with the sheer volume of the data definitions, makes the use of manual Lifecycle Dictionaries impractical. At present, several automated dictionary packages are available. Most of these are being marketed in conjunction with the use and purchase of a database management system. These are based on the physical aspects of data handling, not the logical. They support one particular database management system or another.

The recent influx of personal computers in the realm of business data processing promises to change this situation dramatically. The ability to collect data from a mainframe computer and send data to it is one aspect. Another is that it is now possible to create Lifecycle Dictionaries on a personal computer, maintain them, incorporate the work of several analysts, and reuse the Lifecycle Dictionary portion of a previous project. All of this can be done on a microcomputer, completely independent of a mainframe computer [2]. If necessary, these results can be incorporated into a companywide dictionary. This local, independent Lifecycle Dictionary capability makes accurate, up-to-date Lifecycle Dictionaries a practical reality.

3.6.1 Minimum Features Supported by an Automated Lifecycle Dictionary

The basic features needed by a Lifecycle Dictionary tool are presented in Table 3.6.1-1. This is not intended as an all-inclusive list but, rather, a starting point in developing or selecting one.

TABLE 3.6.1-1:   Minimum Features of an Automated Lifecycle Dictionary

ALPHABETIZED LISTINGS

The system should enable the analyst to obtain a listing in alphabetical order of all data items which appear on the left side of the composition symbol. This listing should be restrictable to a specified set of entries—for example, all entries beginning with a particular character or string of characters.

NEARLY UNRESTRICTED LABEL SIZE

The number of characters that can be used to identify a data item should be large enough to enable the analyst to describe nearly any data item by concatentating name elements together. This is intended to greatly reduce the amount and type of abbreviation that may be necessary.

"WHERE USED" CROSS-REFERENCE CAPABILITY

This provides the analyst with the ability to get a listing in alphabetical order of any or all data items which appear on the right-hand side of the composition symbol, together with the item(s) of which they are a part (i.e., the data item name which appears on the left-hand side of the composition symbol).

"SHORTHAND" RETRIEVAL AND ENTRY

This feature enables the analyst to reference, copy, or modify *any* data item by use of a unique identifier which has been automatically assigned by the system at the time the data item was first entered. For example, a data item like "POLICYHOLDER_ADDRESS_CHANGE_STATUS" is a real pain to have to enter each and every time one wishes to refer to or use it in some way. A unique identifier like CAD0098 is shorter and more convenient. This is particularly true if the three-letter identifier (e.g., CAD) references the project that this dictionary is associated with and it is a default. That is, data from other projects can be referenced or imported by entering the project code. If one is not entered, then the current dictionary is assumed. In such a case, the user would enter 0098.

INTERNAL CONSISTENCY VALIDATION

The system should be capable of:

Identifying any and all data items which appear in a definition but are not defined somewhere in the dictionary

Identifying any and all data items which are defined in the dictionary but which are not referenced somewhere else in the dictionary

Preventing any data item from being entered whose name is identical to some other, preexisting data item

Identifying any and all data items used in the process, information, and/or event models which are not defined or defined as part of those models and not used.

CONFIGURATION CONTROL AND TRACEABILITY

Project(s) used on

Identity of author of each entry, creation date, change authority

AUTOMATIC CROSS-CHECK TO DATAFLOW DIAGRAMS AND STRUCTURE CHARTS

USER-SELECTABLE SUBSETTING OF REPORTS

This would enable the user to select some part or all of the information contained in the Lifecycle Dictionary. The user could, for example, have the system simply print out or display the composition information (which would include any aliasing) *without* any of the other information being presented. Users could select any subset of the information present in the Lifecycle Dictionary, depending on the nature of their problem and the phase of development or maintenance they were in.

REQUIREMENTS TRACEABILITY

Assign each data item to one or more requirements. Relate requirements to themselves as well as to user definable relationship types [2].

SOURCE: Extracted from the seminar, "Structured Analysis for Real-Time Systems," by Software Consultants International, Ltd.; Kent, Washington; Copyright 1988. Reprinted by permission.

3.6.2 A Few Words about Automated Tools

A personal survey of vendors and their products indicates that a lot of ''wishful'' thinking goes on in addition to what might be politely referred to as inaccurate statements. This conclusion closely matches the experience of others who have made inquiries regarding currently available systems or, in some cases, have been unfortunate enough to have purchased one.

As an example of ''misrepresentation,'' some vendors tout the fact that their system enables you to record pseudocode (or structured English, if you prefer) in the dictionary. However, none of them openly admit, unless pressed, that such entries in the dictionary would have to be made in the form of comments. This means that the information contained in those comments cannot be analyzed.

What sort of analysis might one wish to perform on structured English or pseudocode? It would be very helpful, if not essential, that all data items mentioned in the pseudocode or structured English be cross-checked against the dictionary to make sure that they have, in fact, been defined. When pressed on this, the manufacturer of one of the most popular of these employs the RSN (Real Soon Now) strategy by indicating that the desired capability is forthcoming in a new release which will be available in a month. In one case, the vendor held out this promise to a client in October of 1985, and as of April 1986 the feature was still ''forthcoming.''

So what is one little feature, you might ask? Try another. How about a ''state-of-the-art'' system of this type which *cannot* detect duplications of process names, definitions, or data item names and definitions as long as they are more than one level apart in the hierarchy? How about systems which will destroy most or all of your analysis model if you make changes to the top level in your dataflow diagram model? What about one which takes, quite literally, hours to cross-check a thousand data items? How about systems which run on the IBM PC/XT™ or PC/AT™ (Both are trademarks of International Business Machines Corp., New York) but do not use the DOS utility for reading and writing data to disk and, as a result, not only fail often but leave the data in a form which cannot be recovered? How about the fact that many of these systems cannot import or export data? Some systems that are promoted widely have *all* of these shortcomings. Many others have only a subset.

There are a lot more problems with automated aids for use with the structured methods—the list is too long to relate here. A word of advice is in order: remember that the Consumer Protection Act seems to have been suspended when it comes to tools for the structured methods. About the best approach is to be aggressive with salespeople who are trying to sell you a system. Remember that each system's cost of a few thousand to tens of thousands is nothing compared to the costs that problems with them can impose on your project in terms of the most precious resource you have—people's time. As an acid test, tell the salesperson you want to call someone at a firm which is using the system *right now*. Ask that person how many processes, data items, and lines of structured English or pseudocode are involved in the biggest project that they used it on. Find out whether or not they think the system costs more human resource than it saves. Many firms using the most popular of these tools have indicated that the tools actually *cost more than they save*. Also, be wary of demonstrations at trade fairs and the like. The salespeople who are hired to promote these products often know the product well, the methods fairly well, and the product's shortcomings not at all. They do not know the shortcomings because they have not actually used the product *on a real, large, complex* Structured Analysis effort. As a result, they tend to respond to questions in positive terms.

In the author's experience in consulting for organizations, sometimes within the same company, it has happened that one organization would relate problems with a tool which another had never experienced. Further investigation revealed that one organization was dealing with a large-scale effort while the other was dealing with a smaller one or was just getting started and would expand later or was only using the tool on a small part of the project. The makers of the most popular of these tools never had them used on a project equal in size to four out of five of the projects using them that this author encountered as a consultant! As a result, *all* of these software tools failed merely because the number of things that they had to deal with exceeded their limits. In *all* cases, it was the customer's understanding that the tools they were buying could ''handle a project of any size'' or there was ''no practical limit as to the size of project this system can handle.'' CAVEAT EMPTOR!!!

Recently, this situation has changed for the better. Newer automated tools (for example, *EXCELERATOR*™, trademark of Intech Corp., Cambridge, Mass.) provide improvements in some of these areas of concern and one other very important feature—the ability to customize the tool. This enables the user to modify the notation, entities, and objects that are described, and reports to suit a particular project. A set of codes that will modify the notation and other aspects of this tool to correlate with some used in this book is available [4].

The greatest danger in using an automated tool is a failure on the part of the vendor and user to recognize when it is appropriate to use a software tool and when it is not. For example, vendors often envision their product as *the* means by which the user will initially create, refine, and finalize a set of dictionary definitions or diagrams. The problem with this is that the process of creating an analysis is only a process. You begin with what amounts to be a free association of ideas. This is a very creative, unstructured activity that is followed by some degree of enhancement. The end result is validated, revised, and structured into final form.

Most analysts feel that they can generate a dataflow diagram (or other figure related to analysis and design) more quickly on a board than with a software tool. Based on direct observation, they are correct for the early, creative stages of model development. All software tools impose some modicum of structure and discipline. They include syntax and semantics. These factors inhibit creativity. In the early stages of model development, we are attempting to externalize what is private to us. Using a software tool during that stage can prevent us from rapid externalization.

The late stages of model development almost require the use of some software tool. The tool imposes a level of discipline and consistency that is not possible using

manual techniques. Hence, software tools should be viewed as complementing—
not replacing—manual techniques.

The primary benefits of automating the model development process lie in two
areas. One is the ease with which the model can be validated. Systems designers
usually complain about the volume of corrections that must be made when using an
automated system. However, these complaints subside when the second type of
benefit is experienced: This is the ease with which changes can be made. This may
be the most important benefit as we can assure ourselves that changes will always be
necessary.

REFERENCES

1. T. DeMarco, *Structured Analysis and System Specification*. New York: Yourdon Press, 1987.
2. *Lifecycle Management System*. Kent, Washington: Software Consultants International, Ltd., 1987.
3. "Structured Analysis for Real-Time Systems," a seminar by Software Consultants International, Ltd., 1988.
4. "Advanced Structured Analysis and Design Package Level Dictionary," Kent,
   Washington: Software Consultants International, Ltd., 1987.

CHAPTER 4

Entity-Relationship-Attribute Diagrams

We have already examined one way of describing data, the *data* dictionary. It is concerned primarily with describing the hierarchical composition of data. That information is useful in reducing confusion about the meaning of terms and naming of data elements, but it is not sufficient to support the needs of Structured Analysis (and Structured Design). The "missing dimension" is the non-hierarchical relationships that one group of data elements may have to some other group of data elements. For example, a data dictionary definition of an AUTOMOBILE (e.g., year, make, model) would be easy enough to write and refine. So would one for REGISTERED_OWNER and LEGAL_OWNER. The definitions arrived at could be accurate and consistent, but would they be complete? Probably not, unless we somehow note that automobiles, legal owners, and registered owners all have some *relationship* to one another. Instances of automobile are referred to as *objects*. Groups of objects (e.g., automobile) are referred to as *entities*.

Why are relationships so important to consider? The main reason is their potential for being utilized by the enterprise [1] in conducting its business. For example, an automobile dealer may wish to know how many cars an individual is the legal owner of and their description so that he can approach this person about trading in his or her old car for a new one. Another example is the composition of an employee: name, address, telephone number, social security number, marital status, and number of dependents. For a particular object, such as a person named Sally Jones, these elements are related. Sally Jones is an instance of person, and together these elements describe her. There are also other groups of information that are related to Sally—for instance, her employer, type of job, yearly income, and employee identification number.

If we turn our attention to a business, a similar pattern emerges. That is, a body of information is related because it describes the clients of the firm. Another body of information may be related to the products of the firm, and so on. This "enterprise model" is actually the underlying information which constitutes the enterprise. A complete analysis includes the creation of this model of information relationships and not just the hierarchical composition of it. If one already exists for the entire enterprise, then we need only use the parts of it that apply and/or modify that model to include any new information. This information model has several uses, including:

> Creation of a specification from which the database design can be derived. Such a specification cannot be effectively developed from the process model (dataflow diagram).
>
> Improved understanding of the information aspects of the enterprise and system.
>
> Increased possibility of integration and streamlining of the information requirements of the organization.
>
> Identification or correction of undefined or aliased data (via the dictionary).
>
> Identification of causes of data redundancy.

This is in keeping with our earlier discussions regarding what a *complete* analysis model is composed of. We identified three major components:

> Information
>
> Process
>
> Event

Thus far, we have directed our attention at the first of these components. Take another look at our information chart (Figure 4.0-1). Note how the information aspects of the analysis are described by the Entity-Relationship diagram (or E-R diagram). We have chosen to leave out the word ATTRIBUTE, since it was too large, but we will not overlook attributes in this section. The point is that the information portion of the Structured Analysis package is really part of a larger view of information. This larger view we shall refer to as Information Architecture. It is composed of three pieces:

> **Information Model.** This is sometimes referred to as the "Database Model." It consists of the E-R or E-R-A Model, Data Element descriptions, and Subject Databases. These databases are used to support much of the client's business activity.

	DEVELOPMENT PHASE		
TYPE OF MODEL	STRUCTURED ANALYSIS	STRUCTURED DESIGN	IMPLEMENTATION
PROCESS	Dataflow Diagrams Dictionary[1] Pseudocode	Structure Chart Dictionary[2] Pseudocode[4]	Code Dictionary[3]
INFORMATION	Entity-Relationship Diagram Dictionary[1]	Database Design Dictionary[2]	Implemented Database Dictionary[3]
EVENT	Event Model[5] Dictionary[1]	Event Model[6] Dictionary[2]	Queuing Model Dictionary[3]

[1] This dictionary includes data definitions: pseudocode for each process in the analysis model; descriptions, attributes, and content for the entities and relationships in the E–R diagrams; and descriptions of the events and states in the Event Model.

[2] This dictionary contains everything found in the analysis version of the dictionary plus definitions of all flags and pseudocode for all modules that were added as part of the design process or not otherwise present in the analysis.

[3] This dictionary has been updated to incorporate all information developed during the design activity which has been modified as a result of implementation considerations.

[4] The pseudocode referred to here incorporates any and all flags employed in the Structure Chart. All pseudocode is contained in the dictionary as well. Much of this may be identical to or based on the pseudocode that was developed during the analysis.

[5] Event Model refers to a simplified form of state transition diagram.

[6] This Event Model would be further refined and leveled based on design "discoveries."

Figure 4.0-1:   Relationship of Phases to Tools and Model Types [S1]

Business Model. This consists of descriptions of the functions that the business performs, the processes which are used to perform them, and the various policies and activities which surround conduct of the business.

Application Model. This is sometimes referred to as the "Process Model." It describes the systems, programs, modules, and other support and software which enable the business to perform its functions by means of using the information it has.

This architecture is demonstrated graphically in Figure 4.0-2.

The relationship between information modeling activities and the more common view of Structured Analysis activities is shown in Figure 4.0-3. Note how Structured Analysis and Information Modeling are parallel. The primary difference is that they take a significantly different view of systems. As a result, each focuses on only two-thirds of the triangular model presented in Figure 4.0-2. In its more classic form [2], the Structured Analysis activity is concerned primarily with the Business Model and the Application Model, with the Information Model as some-

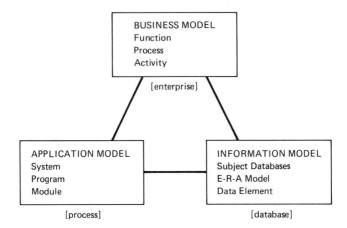

Figure 4.0-2: Information Architecture Overview [S1]

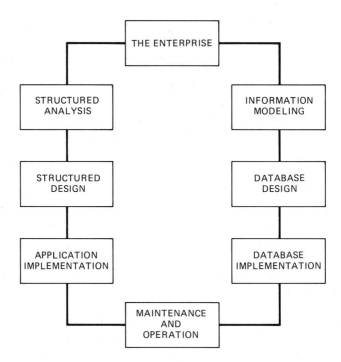

Figure 4.0-3: The Relationship between Information Modeling and Structured Analysis [S1]

thing of an afterthought. Information Modeling focusses attention on the Information Model and either the Business Model or the Application Model, depending on who is conducting the work and what his or her views are.

An alternate way to look at the roles and interactions of the environment and the various parts of the information and database analysis and design activity is presented in Figure 4.0-4.

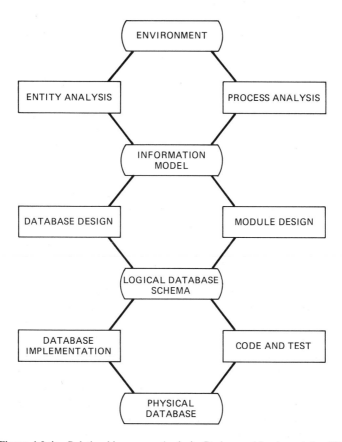

Figure 4.0-4:   Relationships among Analysis, Design, and Implementation [S1]

In this text, we focus attention on the development of a balanced view—one which attempts to keep all three parts of the information architecture in balance. The development of an adequate information model through the appropriate use of E-R diagrams and an expanded form of dictionary is an effective means of doing this.

In this chapter we will examine the process of E-R diagram development as it relates to database specification. This expands the more common view of Structured Analysis as being directed at specifying only the code.

4.1 DATA STRUCTURE DIAGRAMS VS. E-R DIAGRAMS

The software industry's initial love affair with code and programming languages still affects the way we view systems. For example, in Structured Analysis, the primary emphasis in its earlier use was on process models (dataflow diagrams). Data-oriented models were of lesser importance. *Data structure diagrams* were an attempt to describe logical data model requirements. Their main purpose was to provide the analyst with a means of communicating to the database what sort of behavior was expected from it. That's right, behavior! To accomplish this the analyst employs a procedure (described later in this chapter) to attain several goals:

> Identify the minimum set of data from which all data inherent in the enterprise may be derived.
>
> Derive the "natural aggregates" or clusters of data which occur due to the proximity of their use (i.e., they are usually used together).
>
> Ensure correctness and consistency in the description of the database via the data structure diagrams.

As you may recall, when we employed dataflow diagrams, we took a somewhat "laissez-faire" attitude about the data stores that were spelled out on the diagrams. Our basic notion was that they were "temporary" holders of data. The frequency of access, the volume of data, and the coherence of the data (i.e., whether any or all of it was ever accessed, once it was stored) were not major concerns. Now we become concerned about several of these aspects. We are attempting to state to the database designer how the database is to behave. For example, via a data structure diagram, we may wish to communicate the policy that the database must be implemented in such a way that, given the license number of a motor vehicle, the user can obtain the make, model, year, and color of it as well as the legal owner, registered owner(s), and their names and addresses. However, given the name of the registered owner, the system will *not* give the user the name of the legal owner of the vehicle. These rules are based on client interviews and are derived from the analysis dataflow model, data dictionary, structured English, and data stores. Note that the database designer is not restricted as to *how* these rules will be realized in the physical database. The database person is also free to "tune" or adjust the database according to access volumes, response times, or other constraints.

4.2 DATA STRUCTURE DIAGRAM NOTATION

The data structure diagram (DSD) [2] employs a notation similar to that suggested by Bachman [3]. However, there are some conceptual differences between Bachman diagrams and data structure diagrams. The DSD is a network of rectangles and heavy, interconnecting arrows (Figure 4.2-1). These rectangles indicate the exis-

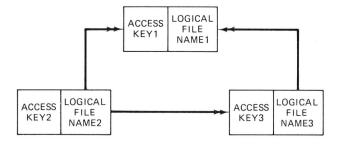

Figure 4.2-1: Generic Example of Data Structure Diagram Notation [S1]

tence of a logical file. If and when the system is automated, each of these may or may not become physical files. That decision is left to the database designers.

Each rectangle is subdivided into two portions. The smaller portion (Figure 4.2-1) is used to indicate how the data is organized within the logical file. That is, it shows the keying or access mechanism. The larger portion contains a label by which the entire logical file may be referred to. The directed line segments which connect the rectangles are derived from an analysis of the system and identification of correlative files. They indicate the sense of the accesses between logical files. For example, the DSD in Figure 4.2-2 implements the policy regarding vehicle registration which we described in Section 4.1

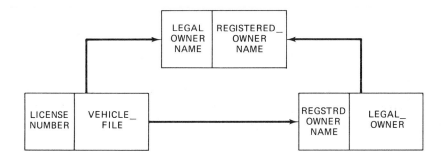

Data dictionary entries for the above might be:

VEHICLE_FILE = LICENSE_NUMBER + MAKE + MODEL + SERIAL_NUMBER + YEAR +
                 NO._OF_CYLINDERS

REGISTERED_OWNER_FILE = REGISTERED_OWNER_NAME + STREET_ADDRESS +
                          CITY + STATE + ZIP

LEGAL_OWNER_FILE = LEGAL_OWNER_NAME + STREET_ADDRESS + CITY + STATE + ZIP

Figure 4.2-2: Example of a Data Structure Diagram [S1]

Note that the vectorlike notation in DSDs documents the existence of a relationship between the "keys" or organizational elements of two logical files. These accessing rules have a directionality. That is, they are either one-way (i.e., a single-headed vector) or two ways (i.e., a double-headed vector).

Many other notations [4] are in use together with the application of a considerable amount of mathematical formalism [5], but these simple, basic tools constitute what is currently used within Structured Analysis.

Notice that the arrangement of the logical file symbols and correlative symbols is not restricted. For example, as long as the correlative symbol extends *from* the logical file intended and *to* the logical file intended, the symbol may touch any part of the logical file symbols. It is also permissible to have correlative symbols cross each other. However, owing to the confusion this may cause, it should be avoided.

4.3 DERIVING DATA STRUCTURE DIAGRAMS

Developing a data structure diagram based on a Structured Analysis model is not as straightforward a task as one is sometimes led to believe [2]. The early stages are the most critical to the success of the effort, while the later ones are so close to being mechanical as to be prime candidates for automation.

Given that we have developed the current physical model via leveled, balanced dataflow diagrams with accompanying data dictionary and structured English statements, we can employ a multistep process to derive a DSD:

1. Catalogue all physical data accesses

This is accomplished by noting which data items go into each datastore and which come out. At this point, these data items may very well be such that they can be referred to with an aggregate label such as CLIENT_CONTACT_DATA. We may encounter data elements or items which never leave a datastore or whose origin is unknown. These will require further investigation to resolve inconsistencies. We also need to note whether a data item is read or written (or both) and by which process or processes the above is done at all levels in our dataflow diagrams.

2. Produce the minimum data model (through inference)

Our reason for documenting which processes utilized the data was so that we could examine what the data was used for. For example, it is not only possible but common for one process to access data in order to produce some sort of summary or analysis when, in fact, that result exists in another datastore. Remember, the system we are analyzing is the result of a "hit-and-miss" evolutionary process—*not* some rational, logical progression toward an end result. Independent, asynchronous data acquisition and use have taken place over a period of years. We may also expect that when data items are identified for which we cannot find any use, there will be resistance to eliminating them.

A colleague at a Southern California aerospace manufacturing firm ran into this latter problem a few years ago. He and his team had identified a rather significant volume of data associated with something called a VENDOR_PERFORMANCE_INDEX. They could not identify anyone in the division or corporate structure who used it or the data elements that went into it. They suggested eliminating it and its database, which produced an outcry about how important it was. The issue was finally resolved by instituting a detailed data storage accounting system which charged each group that "owned" a certain data set a fee for storing it. Since no one would claim ownership (and cost responsibility) for this data, it was recorded on several reels of magnetic tape and removed from the database. Do not be surprised if something similar happens to you and you have to resort to some sort of contrivance in order to bring logic to bear on an illogical situation.

3. Package the Results for the Database Designer

This step involves eliminating any and all data elements which can be derived from other data elements. It borders on heresy with respect to some notions of "efficiency." Remember, the result of the process (a data structure diagram) is a statement of the relationships and access rules the (eventual) implemented database must support. It is *not* a statement of how it is to be implemented. The database designer is free to implement in any way deemed prudent.

The three steps are depicted in more detail in Figure 4.3-1.

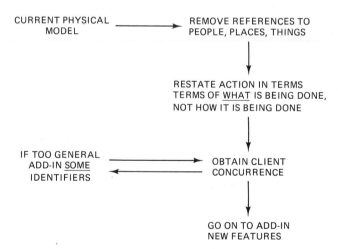

Figure 4.3-1:   Logicalization to Produce DSDs

4.4 DATA STRUCTURE DIAGRAMS VS. DATABASE SPECIFICATION

The most important thing to keep in mind regarding data structure diagrams is that they describe how the database is to behave and *not* how it is to be physically implemented. Stated another way, DSDs are intended to portray the (eventual) database from the following standpoint: it should be constructed in such a way as to behave *as though it were constructed* this way. Again, the database designer is free to implement the database in just about any manner deemed appropriate.

For example, in Figure 4.2-2, where there is a single, direct access between two logical files (e.g., between REGISTERED_OWNER and VEHICLE_FILE), this may be implemented physically as multiple accesses. Similarly, what appears in Figure 4.2-2 as a single logical file may be implemented physically as several files. There may be several good reasons for doing this at implementation time, including the economics of restructuring a large, existing database, the timeflow required to bring up the new database, and the statistics associated with the use of a portion of the database. For example, multiple accesses may be quite justified if we are dealing with one or more events which occur infrequently.

The point to keep in mind here is that a data structure diagram is an expression of how the database will *appear* to be physically organized, *not* an expression of how it *is* organized. The distinction lies in the fundamental differences in viewpoint between what is conceptually possible and what is practical and prudent. The fact that DSDs do not contain any data that can be derived is an important consideration for the database designer but not a *primary* consideration during analysis.

4.5 SHORTCOMINGS OF DATA STRUCTURE DIAGRAMS

Although the DSD represents an important shift in our perception of what should/should not be included in the analysis activity, it still leaves much to be desired. It has three serious shortcomings:

- It includes implementation considerations, since it clearly depicts the key or organizational structure within each of what it calls Logical Files. This tends to bias the implementation.
- The process by which the DSD is derived is difficult and uncertain. Success requires an incredible amount of insight and foresight, particularly in the early stages.
- Its relationship to the other aspects of the analysis model (i.e., event and process) is not strong.

Hence, our search for an effective means of describing the information relationships is not met by DSDs. We still need a means of describing information relationships from a purely logical point of view.

4.6 ENTITY-RELATIONSHIP CONCEPTS

Entity-Relationship diagrams provide precisely what we are looking for to represent information relationships. They give us an effective means of capturing the "clusters" or "aggregates" of information and the relationships these have to each other.

These relationships are a direct result of the nature of the enterprise or business activity which we are modeling. This so-called information modeling approach is part of an overall redirection of our image of what analysis is and what it should include. These diagrams can take many forms. The Entity-Relationship (E-R) diagrams that we shall be using depict two kinds of information: entities and relationships.

4.6.1 Entities

An *entity* is an object or a group of objects. An *object* is a "thing" (e.g., person, place, organization, function) about which we wish to collect information. It is said to have a significant role, purpose, function, and/or other characteristics within the system being modeled. An object always has a singular name. Although the details about objects (e.g., automobiles, televisions, radios) may vary from one member to another, they share certain things in common. It is this very commonality which we wish to take advantage of in both our modeling and our (eventual) database implementation.

4.6.2 Relationships

A *relationship* is a significant association or interaction involving one or more entities. A relationship depends wholly upon the existence of entities for its own existence. It cannot stand alone.

Together, the entities and relationships form a network (graphically) which describes that portion of the enterprise which we wish to model. Vital to the creation and use of these diagrams is an accurate, consistent, concise, and complete Lifecycle Dictionary.

4.6.3 Notation for E-R Diagrams

E-R diagrams are often composed of rectangular and diamond-shaped symbols connected together by straight lines to form a network. Labeled rectangles are used to represent entities. The label is the name of the entity. We have chosen to use angular brackets (which can be depicted by a pair of "greater than" and "less than" symbols) to represent relationships. In cases where either/or relationships exist, a combination of intersecting lines with a fan-out are used.

Numbers are used on the connecting lines to indicate the relative rate of occurrence of objects with respect to other objects via the relationship indicated. This relative-rate-of-occurrence property (i.e., the rate of occurrence of one object to one or more other objects) is known as *cardinality*.

Relationships between objects can be read in either direction if it is consistent with the nature of the system involved. In many cases, it does not make sense to have a reversible relationship. For example, one PROJECT_MANAGER is said to MANAGE one or more PROJECTS. However, it does not appear to make sense to say that one or more PROJECTS MANAGE a PROJECT_MANAGER (even though the staff of those projects may feel that they are managing the PROJECT_ MANAGER). Hence, we must have some means of documenting this fact. In this case, we indicate that PROJECT_MANAGER is an *anchor* point. A bracketed number indicates that the relationship is such that it must be read in a "from—to" sense. That is, in reading the relationship, we start with the object nearest the bracketed number (the anchor object) and end with the object nearest the un-bracketed number. The notational elements are described in Figure 4.6.3-1.

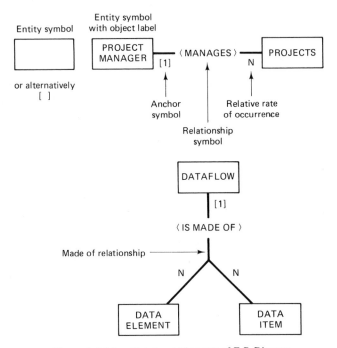

Figure 4.6.3-1:   Notational Elements of E-R Diagram

For example, in Figure 4.6.3-2 the relationship between REQUISITION and PURCHASE ORDER is called "CREATES." According to the diagram, "[one] REQUISITION CREATES /one/ PURCHASE ORDER." This is consistent with the directional "sense" indicated in the diagram. To read the relationship in the opposite direction (i.e., backward) not only violates the "sense" of the diagram but does not seem reasonable. That is, /one/ PURCHASE ORDER CREATES [one] REQUISITION" seems inconsistent with reason as well as with the diagram.

The key to understanding the role of E-R diagrams lies in the nature of relationships. Relationships are an inherent part of a software system. They are often *unique* to that enterprise. For example, E-R diagrams of two different com-mercial aircrafts may have many of the same objects (such as wing, engine, on-board computers) but their relationships may be unique to each manufacturer.

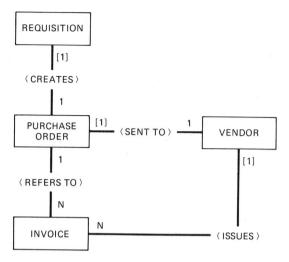

Figure 4.6.3-2:   Example of E-R Diagram Notation [S1]

4.7 STANDARDS FOR USE WITH ENTITY-RELATIONSHIP DIAGRAMS [S1]

The standards take two forms:

> Conceptual
> Notational

Each is discussed below, together with some exceptions.

4.7.1 Conceptual Standards for E-R-A Models

1. An entity model is a conceptual representation of real-world, enterprise objects.
2. Entity models are composed of entities and relationships.
3. There may be several entity models for each subject database.
4. The entity model contains objects of interest to functions and processes of the external world.
5. The entity model frequently contains information that is needed for completeness, rather than information that is useful only to the current application(s).
6. Relationships are associations between entities.
7. Each relationship has a name, and one or more initial and terminal entities.

8. Relationships are like entities in that they represent types rather than actual instances.

9. Actual instances of a relationship may have differing numbers of initial, terminal entity tuples. The number of terminal entity objects may also vary.

10. Relationships can be 1-to-1, 1-to-many (1-to-N), or many to many (N-to-M). For example: 'Lives-in' is a relationship between person and house. Since a person may live in 0 or more houses, and 0 or more persons may live in a house, it is N-to-M. If 'house' means primary residence for tax purposes, it is N-to-1. (Note that the two Ns may not represent the same number!)

11. Entities may be related to other entities by means of one-to-one, many-to-one, or many-to-many relationships.

4.7.2 Notational Standards for E-R Diagrams

1. Each entity is represented by a rectangle.

2. Each relationship is represented by a pair of angular brackets (greater-than and less-than symbols).

3. If a relationship is too long to be conveniently fitted on a single line, two lines may be used, with each surrounded by its own pair of angular brackets.

4. Anchor objects are indicated, wherever deemed appropriate and consistent with the nature of the system being modelled.

5. Entities are labeled using a descriptive noun.

6. Adjectives are not to be used in describing entities.

7. Relationships are labeled by means of a verb.

8. Adverbs are not to be used to describe relationships.

9. All entities and all relationships are to be defined in the Lifecycle Dictionary.

4.8 DERIVING E-R DIAGRAMS [S1]

The process of deriving a first cut set of Entity-Relationship diagrams is a rather simple one although it does not directly involve the Business model, its use of the Process model is assumed to accomplish this link. The derivation process requires the development of a level-'0' dataflow diagram. It is recommended that the level-'0' diagram first be validated by developing level-'1'. The steps leading to an E-R diagram are described below:

1. Examine the level-'0' dataflow diagram for the system and identify the dataflows between processes and those between sources/sinks and processes, the datastores present in the diagram, and the sources/sinks from which the data emanates. List these.

2. For each item listed in step 1, determine whether or not it provides us with new information and whether or not we wish to keep that information.

3. Form a network of entity and relationship symbols which incorporates those items in step 2 which provided *both* new information and that which we wish to keep. Relationships are to be labeled with a singular verb. Add in entities, as necessary.

4. For each pair of entities separated by a relationship, assign a 1 or an *N* to represent the nature of that relationship—that is, whether or not there will only be one instance or many instances of one entity for each instance of the other. Each relationship which results in an *N*-to-*N* relationship, should be decomposed into two or more relationships which are 1-to-*N*. *This may require the application of this process to the next lower level dataflow diagram.*

5. Check results for completeness by deriving the data necessary to support each and every process on the level-'0' diagram that we started with. If there is any needed information which cannot be derived, add whatever entity(s) may be necessary. Recheck, as necessary.

6. Validate results by listing the entities on the diagram. Identify whether or not we truly wish to keep data on them.

4.9 EXAMPLE OF E-R DIAGRAM DERIVATION [S1]

To show how E-R diagrams can be derived employing the approach described above, we will use the following simple example.

> A company has a system in which departments can obtain the services of qualified vendors through the use of purchase orders. In order to do this, they must submit a requisition for approval for each vendor to be employed. Approved requisitions result in the generation of a purchase order. Each purchase order is communicated to a specific vendor. Upon completion of the work, the vendor submits an invoice in order to request payment for the work.

Analyzing this statement, we generate the dataflow diagram (using rectangles for processes) presented in Figure 4.9-1. We may now proceed to execute the steps described above.

Step 1

The items for our list are:

REQUISITION
APPROVED_REQUISITION

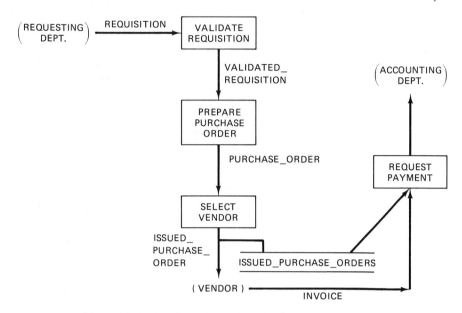

Figure 4.9-1:   Dataflow Diagram for E-R Diagram Derivation [S1]

PURCHASE_ORDER
VENDOR
ISSUED_PURCHASE_ORDER
INVOICE
APPROVED_INVOICE
REQUESTING_DEPARTMENT
ACCOUNTING_DEPARTMENT

Step 2

NAME	NEW INFORMATION/KEEP?
REQUISITION	Yes/Yes
APPROVED_REQUISITION	No/No
PURCHASE_ORDER	Yes/Yes
VENDOR	Yes/Yes
ISSUED_PURCHASE_ORDER	No/No
INVOICE	Yes/Yes
APPROVED_INVOICE	No/No
REQUESTING_DEPARTMENT	Yes/No
ACCOUNTING_DEPARTMENT	Yes/No

Steps 3 and 4

The E-R diagram resulting from these steps is presented in Figure 4.9-2.

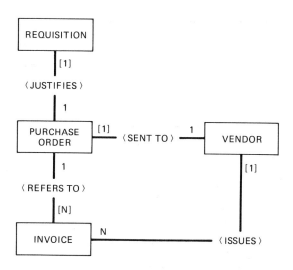

Figure 4.9-2:   Resulting E-R Diagram [S1]

Step 5

The validation of results is demonstrated below:

Name	What It Is	Keep Data on It?
Requisition	Thing	Yes
Purchase_Order	Thing	Yes
Vendor	Person	Yes
Invoice	Thing	Yes

4.10 DESCRIBING ENTITIES IN THE DICTIONARY

Entity-Relationship diagrams document the relationships which exist be-tween/among the groups of objects which are part of the system being studied. Although this graphic depiction is an efficient and compact means of describing an essential part of the database specification, it is not complete. Many properties of the entities and relationships are *not* described on the E-R diagram. For example, an entity may have a unique identifier by which we can single out a given instance of it. Relationships have a similar problem in that they may involve policies which cannot be stated on the diagram.

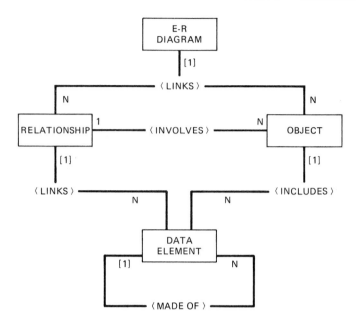

Figure 4.10-1:   E-R Diagram of E-R Diagrams [S1]

The E-R diagram presented in Figure 4.10-1 will be used to direct the discussions presented in the rest of this section.

Figure 4.10-1 tells us that an E-R diagram may be documented by describing relationships and objects. The properties that these have will be described in more detail below through the use of an example problem. The problem or system that we will be discussing is the seminar organization within a consulting firm. This resembles the classic example used in other texts. However, we have simplified it here by relating it to a simpler situation. Figure 4.10-2 describes the relationships and entities that exist within our example. Note that INSTRUCTOR is an example of an entity with two anchor points. With this example we will demonstrate the role of the E-R diagram as part of the Information Model and the Relationship between it and the Lifecycle Dictionary.

The documentation of the objects and relationships described in Figure 4.10-2 is discussed in the sections presented below. As a *minimum,* the definition of each data element would be composed of the topics listed in Table 4.10-1.

TABLE 4.10-1:   Topics Required to Describe
Objects and Relationships

NAME
CONTENT
STRUCTURE
DEPENDENCIES

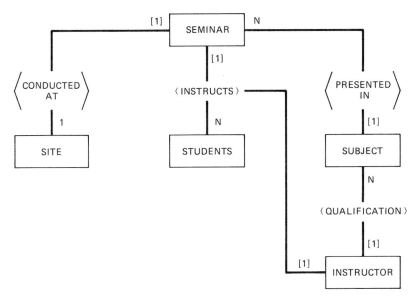

Figure 4.10-2:   Example E-R Diagram to Demonstrate Definitions [S1]

4.10.1 Documenting Objects

One of the entities presented in Figure 4.10-2 is SEMINAR. Its description in the form described above is presented in Figure 4.10.1-1. The information presented in that figure would be recorded in the Lifecycle Dictionary.

Let's take a closer look at the information presented in Figure 4.10.1-1. Notice that under the section marked Identifier, a Unique_id is listed. That identification is presented as

```
SUBJECT_NAME + RELEASE_NUMBER
```

Some readers may ask why we did not make subject_name an object and release_ number an object. The answer is that we need to have a means of uniquely identifying each member of the object, SEMINAR. Also, if we made these individual objects, they would be related to the object SEMINAR by means of relationships with names like HAS. Such "fuzzy" relationships imply that one or more of the objects involved are not objects at all. In addition, subject_name seems a poor candidate as a descriptor for a class of objects rather than a property or characteristic of another object. Hence, we have chosen to associate subject_name + release_ number with the object SEMINAR as an identifier which is unique. Attributes were selected in a similar manner. However, it is possible that the seminar_title, for example, could have been included in the unique identifier. Why did we not include it? The reason is that to do so would be inconsistent with the way in which the enterprise has historically identified its SEMINARs.

NAME: SEMINAR

DESCRIPTION:

A seminar defines the method of instruction in some software engineering area.

A seminar consists of teaching materials, student materials, and (optional) a consulting report.

A seminar is taught to a class by a qualified instructor.

A seminar is a product sold to a client.

CONTENT:

SUBJECT_NAME

SEMINAR_TITLE

SEMINAR_DESCRIPTION

FEE

PREREQUISITE

TEXTBOOKS

RELEASE_NUMBER

VERSION

STRUCTURE:

Identifier: SEMINAR
        Unique_id: subject_name + release_number
        Attributes: seminar_title
                    seminar_description
                    fee
Attribute: PREREQUISITES
        Sub-attribute: prerequisite_name
Attribute: TEXTBOOKS
        Sub-attribute: text_title
                            text_author
                            text_publisher

Figure 4.10.1-1:   Example of Object Description [S1]

From the foregoing we see that the structure and content of the information model are functions of both the nature of the enterprise *and* our use of the technology. In this way we are, essentially, letting the data tell us what the relationships and descriptions are rather than imposing our own views. Certainly, it may happen that we identify some idiosyncrasies in how the enterprise is organized from an information standpoint. These should be brought to the attention of the client but do not mean that there is something wrong with the enterprise. This is just the way in which it works. This is why the information models done for four different savings and loans may not be identical, even though they are governed by the same set of laws, compete in the same areas of commerce, and are of the same size.

4.10.2 Documenting Relationships

The description of a relationship following the guidelines set forth in Table 4.10-1 is presented in Figure 4.10.2-1.

NAME: QUALIFICATION

DESCRIPTION:

Basis:

               For each INSTRUCTOR, there is:
                  at least one subject area

               An INSTRUCTOR is qualified to teach a seminar if:
                  The INSTRUCTOR has team-taught the seminar at least twice
                  OR
                  The INSTRUCTOR is deemed qualified by the Technical Director
                  and has been approved by the client

Binary Associations:

               Each SEMINAR is associated with one or more INSTRUCTORS

CONTENT:

               instructor_name
               subject_name
               seminar_revision
               qualification_date

STRUCTURE:

Identifier: QUALIFICATION
        Unique_id: instructor_name + subject_name + seminar_revision + qualification_date
Attribute:
        Unique_id: none

Figure 4.10.2-1:   Description of a Relationship [S1]

4.11 SUMMARY

The E-R diagram and accompanying dictionary entires will be utilized in the Structured Design phase (Chapter 11). We will use it to identify the accesses to the database, the ramifications for the physical design, and the interplay between the database design and application or code design.

Even without such applications, the E-R model provides us with invaluable insights into how objects of interest interact. Often, clients find that the mere act of defining and relating this information has the effect of helping them to ''see'' their business in new way. These new insights often lead to further inquiries and potential changes of policy.

REFERENCES

1. P. Chen, *The Entity-Relationship Approach to Logical Data Base Design.* The Q.E.D. Monograph Series on Data Base Management, No. 6. Wellesley, Mass.: Q.E.D. Information Sciences, Inc., 1977.

2. T. DeMarco, *Structured Analysis and System Specification.* New York: Yourdon Press, 1978.

3. C. W. Bachman, "Data Structure Diagrams," *Data Base,* The Quarterly Newsletter of the Special Interest Group on Business Data Processing of the ACM, Vol. 1, No. 2 (Summer 1969), pp. 4–10.

4. M. Flavin, *An Introduction to Information Modeling.* New York: Yourdon Press, 1981.

5. C. J. Date, *Introduction to Database Systems.* Massachusetts: Addison-Wesley, 1981.

S1. Extracted from the seminar, "Structured Analysis for Real-Time Systems Seminar," by Software Consultants International, Ltd.; Kent, Washington; Copyright 1988. Reprinted by permission.

CHAPTER 5

Dataflow Diagrams

When we examine the problem of describing a system, we find that the sheer volume of information that we must deal with can be overwhelming. A compact way to describe just about anything is by using graphics. But the use of graphics in analysis has at least two drawbacks.

One drawback is the complexity of the system being analyzed. It is not intuitively obvious how one would describe a complicated system graphically. This is true even though graphics are an extremely powerful means of abstracting and documenting complex systems. One way in which this issue can be addressed is through the use of hierarchy or abstraction.

The other drawback is just as formidable. It is the fact that the system we are attempting to depict graphically is not a set of sequential processes but an interrelated network of asynchronous processes. Effectively addressing this one is not as easy. One effective approach would be to employ the same kind of graphic tools used by engineers who have tried to lay out graphic portrayals of construction projects. It may not be obvious when we observe one, but such projects are an orchestrated set of asynchronous activities. These plans involve many people, tasks, and activities. Examples of the graphic tools that are used include PERT Charts/Networks and Activity Networks. In Structured Analysis, we employ very similar concepts but tailor the notation to make it more suitable for use in software-related analysis efforts.

The single graphic element that is both the most similar to other graphic approaches in engineering and the most widely recognized as being part of Structured Analysis is the dataflow diagram (DFD). It lies at the very heart of Structured Analysis. Its use encourages—even requires—that the analyst view the system being studied in a precise way. The dataflow diagram utilizes concepts that were incorporated into other fields quite some time ago. It bears a strong resemblance to circuit diagrams and production line models of the past, when vacuum-tube circuitry and industrial engineering notation were in vogue.

The dataflow diagram is not unique, but its use in conjunction with complementary tools and the effect those tools have on the software design phase may be. DFDs expand upon the notion of *I*nput-*P*rocess-*O*utput or IPO. This concept is not new. It has been employed for several hundred years in the analysis of heat-transfer problems. It is also sometimes referred to as an energy-balance analysis. It was more widely developed for use in the software field by IBM and others [1, 2], who were originally responsible for the publication of literature and a corresponding graphics template for use in drawing an associated set of graphics.

The basic idea behind the IPO concept or HIPO (Hierarchical IPO) is to look at a system as having one or more inputs, performing some specific process or combination of processes on each class of input data, and having one or more outputs. Again, there is a strong similarity to the field of engineering. The graphics in particular are a lot like those in electrical engineering. For example, the symbol for a datastore in Structured Analysis is nearly the same as the one used by electrical engineers to represent a capacitor—a device for storing electric charge. In that field, the concept of a "black box" is employed in much the same way as the concept of a process in Structured Analysis. Here the concept is implemented via a network-oriented modeling approach which accounts for the source of each piece of information entering the system, the destination of each piece of information leaving, and the transformation or process which changed the input into the output. These transformations can be depicted at any convenient level of detail. The data which appears on the dataflow diagrams is catalogued and defined in the Lifecycle Dictionary (Chapter 3).

5.1 DATAFLOW DIAGRAM NOTATION

Dataflow diagram (DFD) notation is simple and easy to use. It entails few restrictions regarding form and content. This increases its utility, because, beyond these few restrictions, the analyst is free to represent the system in any way that seems appropriate.

The notational requirements are presented below for each of the four types of graphic elements which comprise dataflow diagrams:

Sources and Sinks of data. These represent the people, places, or things that are providers or consumers of information with respect to the system. They are not part of the analysis. Some authors refer to them as terminators.

Dataflows. These are the paths or conduits by which information is conducted from one part of the system to another.

Transformations or Processes. These modify or change data from one form into another.

Datastores. These are temporary holders of data.

We will use two forms of notation. One is similar to the original DFD notation and will require graphics capability to automate on a computer. The other is a derivative form and can be automated without graphics capability. Readers are free to choose which will work best in their own situation.

Be forewarned that most of the automated systems available today to aid in the use of Structured Analysis have treated the more complex type of notation as being sacrosanct. It is not. The notation is intended to provide us with a means of describing the model. In other words, the dataflow diagram is only one of several means to an end—a quality system specification. It is not an end in itself! It is the analytic and consistency characteristics of the diagrams that make them useful. They must also be easy to change without destroying earlier work. This is a shortcoming of many of the automated tools. They seem to have focused their attention on the pictures rather than their meaning and their basis.

5.1.1 Dataflows

For those of you whose parents wanted you to become a plumber instead of a systems analyst, take heart! Structured Analysis and plumbing have a lot in common. (To those of you who think plumbing is the pits, try surviving without it and/or compare your wages with those of a plumber.)

Dataflow diagrams are a lot like the diagrams that plumbers use in laying out new construction. They show input, output, and a variety of routes and devices that are involved in transforming these flows. In the basements of large buildings, for example, we find that the pipes which carry a variety of material such as hot water, cold water, and air have labels on them. These labels usually include an arrow which indicates the direction of flow of the material inside the pipe. Similarly, in Structured Analysis, we use pipelines which have many of the same characteristics. The characteristics of dataflows are described below:

They do not leak—data is neither lost nor gained within the confines of these pipelines. However, unlike physical pipes, they have no volumetric limits. That is, they have unlimited capacity.

They do not change any of the data that passes through them.

Dataflows may be *empty*—just like plumbing lines. The pipe marked "COLD WATER" may be empty. The label tells us that if there were anything in the pipe, it would be cold water.

They do not carry anything but information that is inherent to the enterprise. Specifically, they do not carry any flags, signals, or indicators of conditions.

Dataflows usually emanate from or are directed to processes (Section 5.1.2). In some cases, dataflows are intended to carry data to two processes. That is, the *same* data is intended to go to some set of processes. To show this, we would use the fan-out notation shown in Figure 5.1.1-1. There is another situation involving more than one dataflow from a process. It is one in which data may be present in one or more of the dataflows but not necessarily in all of them. The notation for this situation is demonstrated in Figure 5.1.1-2.

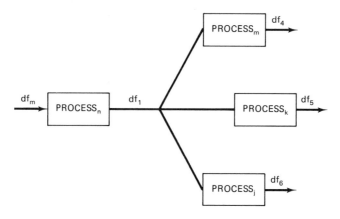

Figure 5.1.1-1:   Notation to Show Dataflow Fan-Out

Figure 5.1.1-2:   Notation for Dataflows Selected by a Process

5.1.2 Transformations or Processes

Transformations do just what their name implies—they change data. By change, we do *not* mean that a transformation (also known as a process) creates something out of nothing. Rather, transformations often involve changing the form or the content of data. One rule that we can state about the use of transformations or processes (sometimes referred to as ''bubbles'') is:

Data may be processed in either of two ways: physically or logically.

A generic example of the notation used to describe processes is presented in Figure 5.1.2-1. Each of the two types of transformation is discussed in more detail below.

Use a circle, a square, or a rectangle to enclose a <u>brief</u> statement of
<u>WHAT</u> the process does, not how it does it

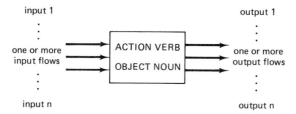

An "ACTION VERB" is a transitive or strong verb

The "OBJECT NOUN" is the receiver of the action

For example, a process with a label like, "VALIDATE_ACCOUNT_#"
has an action verb VALIDATE and an object noun ACCOUNT_#.
The noun is affected by the change.

Figure 5.1.2-1:   Notation Used to Describe Processes

5.1.2.1 Physical Transformation

Physical transformations change the input data in such a way that it is no
longer recognizable. An example is a process which takes place at your local fast
food restaurant every day:

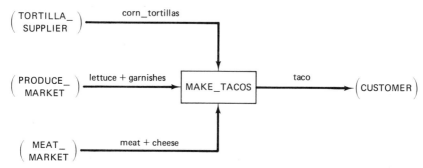

Another example is a process which takes all of the individual transactions
from the day's sales activities and sums them up to give us the total business activity
for that day:

Note that although the Daily Total did not exist earlier, it was derivable from the
data provided to the processor by means of the application of some procedural rule

or mathematical formula. In such a transformation, there is an obvious difference between what was input into the processor and what came out of the processor.

5.1.2.2 Logical Transformations

A process which *logically* transforms data does not change it physically. That is, the use of the data is affected by the way in which the process classifies it and *not* by any physical change. An example is a process which validates charge account numbers:

Note that the actual data that entered the above "bubble" or process is physically identical to the data that left it. Imagine drilling three holes—a hole in the pipe that enters the process, a hole in the "valid" pipe, and a hole in the reject pipe. If we looked into each one, we would find that the physical characteristics of the information in those pipes is the same. So how has this data been transformed? It has been transformed in terms of the character or nature of the data contained in the VALID pipeline. That is, each and every data element found to be in that pipeline will be further processed differently than data found in the CANDIDATE or REJECT pipelines. This data has not been transformed physically. It has been transformed logically.

5.1.2.3 Labeling Processes

One problem that many analysts have with the use of dataflow diagrams is in the area of process naming. The problem lies in the level of knowledge that the analyst has acquired regarding the system being studied. If we are unsure of what we are dealing with, we may tend to employ process labels like "Process_Client_Data," where we are not really sure just what is meant by either Process or Client Data. In Structured Analysis, the emphasis is on accuracy, completeness, and consistency. Hence, it is recommended that certain process-naming conventions be observed. The basic rule is:

Always use a strong (transitive or "action") verb and an object noun to label processes in dataflow diagrams.

A corollary to the above rule is

If you cannot create a name that has the recommended characteristics, leave the process name blank. Someone else will fill it in later, or it may be clearer later just what an appropriate name is.

Examples of appropriate and inappropriate verbs and object nouns are presented in Table 5.1.2.3-1.

TABLE 5.1.2.3-1: Process Labeling Examples

VERBS	DATA NAMES
These are OK: VALIDATE COMPUTER CALCULATE	*These are OK:* ACCOUNT-# TARGET COORDINATES
Take care with these: CREATE EDIT DISTRIBUTE PRINT PRODUCE UPDATE	*Take care with these:* RECORD FILE
Avoid these: PROCESS	*Avoid these:* DATA INFORMATION

5.1.3 Terminators

Terminators are used in Structured Analysis to, in a sense, anchor the dataflows. They are used to identify the sources of data and the sinks or destinations to which the data eventually goes. They provide the analyst with a graphic means of identifying just what the sources and destinations are that are associated with a particular study. These include, but are not limited to, outside departments, people, agencies, companies, government agencies, or other legal entities.

The main purpose for terminators is to ensure that we do not have data in the system under study that cannot be accounted for and that all data has some defined destination. Terminators are also referred to as sources or sinks, a terminology

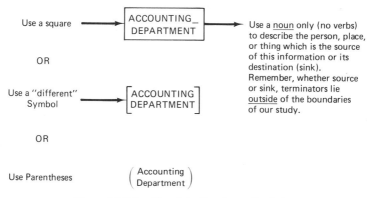

Figure 5.1.3-1:   Use of the Terminator Symbol

borrowed from the scientific field. There, conservation of energy, for example, dictates that there can be no unaccounted-for sources of energy or consumption of energy (sinks). DFDs have the same requirement. The system being studied is treated in much the same way that scientists treat physical systems. In our case, since the outside sources and sinks of information are not part of our study, we do not care how the data is generated nor what is done with it. This concern will change during the design phase. During analysis, we merely need to account for it. The use of these symbols is demonstrated in Figure 5.1.3-1.

5.1.4 Datastores

Datastores meet the need for many systems to store information for later use. They are *temporary* storage facilities for information and are not intended to be permanent.

Datastores cannot change the data that they are holding. What goes in is exactly what comes out. However, when a process accesses data contained in a datastore, it may access a subset (i.e., certain data elements that are part of a larger set of data elements).

The rules regarding datastores are:

One or more elements in a datastore may be accessed, as necessary.

Datastores do not change any of the data left in their care.

Data that enters a datastore must eventually leave that datastore.

The accessing (e.g., keys) mechanism is not depicted or documented.

An example of a datastore notation is presented in Figure 5.1.4-1.

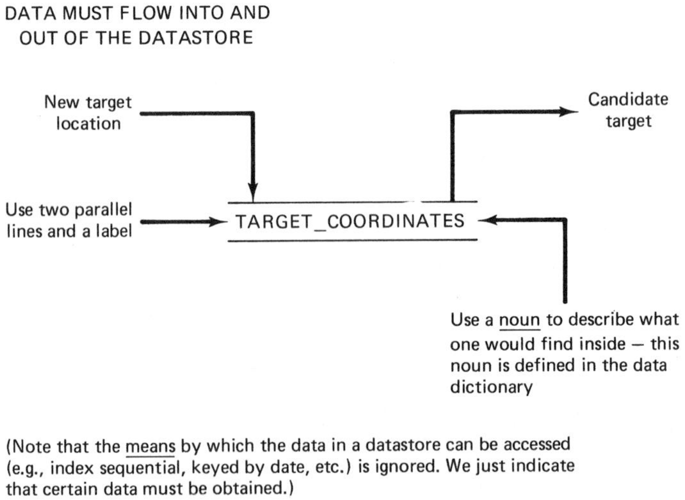

(Note that the means by which the data in a datastore can be accessed (e.g., index sequential, keyed by date, etc.) is ignored. We just indicate that certain data must be obtained.)

Figure 5.1.4-1: Use of Datastore Notation

5.2 STANDARDS AND GUIDELINES FOR USE WITH DATAFLOW DIAGRAMS

In this context, the terms *standard* and *guideline* are used interchangeably. Both are treated as advisories. There are two types of standards:

Notational. These refer to the mechanics of what is drawn.

Conceptual. These refer to the basis for drawing the dataflow diagram.

Each of these together with exceptions is listed below and presented with the associated elements of the DFD.

5.2.1 Standards for Dataflows

5.2.1.1 Conceptual standards for dataflows

Dataflows should be thought of as being pipelines which conduct information from one process, sink, source, or datastore to another.

Dataflows cannot change or modify data in any way.

Data cannot be created or destroyed within a dataflow.

Dataflows cannot be used to conduct flags or signals of a condition.

There will be conditions or instances where a given dataflow is empty.

Dataflows may not pass *directly* from a terminator to a datastore.

Dataflows may not pass from one terminator to another.

5.2.1.2 Notational standards for dataflows

A dataflow is represented by a directed line segment with a label.

Each dataflow must be labeled.

The label must be a noun, preferably without adjectives.

No verbs are to be used in naming dataflows.

All of the names on the dataflows must appear in the data dictionary and be defined there.

Dataflow names should not contain the words DATA, INPUT, OUTPUT, or other "blanket" or generic terms.

5.2.1.3 Exceptions to the dataflow notational standards

A dataflow need not be labeled if, in the opinion of the software engineer involved, labeling it would do more harm than good with respect to communication among other project members. Examples include having a dataflow from a datastore wherein the dataflow's content is apparent. The "label everything" approach can badly clutter the DFD and reduce communication.

5.2.2 Processes

5.2.2.1 Conceptual standards for processes

A process transforms or changes data.

Processes can neither create nor destroy data (i.e., all output from a process must be derivable from the input using some procedure).

Processes receive only the data that they use (i.e., no data is to be "passed through" a process without being changed in some way).

Processes can change or transform data in either of two ways:

Physically. The input bears little or no resemblance to the output, as in the case of sending a mortgage amount, interest rate, and payback period to a process and having the amount of the payment output.

Logically. The input and output are physically the same but are treated differently, as in the case of a process which checks the validity of account numbers according to some rule. The input and output are identical physically, but the output will be treated differently than the input, since the system has established that the account number has some property which the input dataflow could not be assumed to possess.

Processes are like vacuums in they "draw" data toward themselves through the dataflows. Processes *do not* "send" or push data down the dataflow pipelines. They should be thought of as placing data at an exit port which another process can draw by using a dataflow "pipeline."

The external "trigger" which causes a given process to begin or cease "drawing" data and transforming it may be present in the system without necessarily being indicated on the dataflow diagram. (This depends, in part, on which form of the dataflow diagram notation one subscribes to.)

A process is required to stand between or intercept any information flowing from or going to a terminator or datastore.

5.2.2.2 Notational standards for processes

A process may be represented by a circle or a rectangle containing the label of the process and an identifying number.

The label of a process is composed of a strong or transitive verb and an object noun.

Process labels do *not* contain adjectives or adverbs.

A process's identifying number is a concatenation of the identifying numbers of its hierarchical ancestors with a local identifier. For example, the four processes which represent the decomposition of a process with an identifying number 1.2 would have labels 1.2.1, 1.2.2, 1.2.3, and 1.2.4. Similarly, the label 1.2 tells the reader that this process is subordinated to process 1.0.

Process labels should not contain the words PROCESS, EDIT, DATA, IN-FORMATION, GET, READ, PUT, WRITE, STORE, CHECK

5.2.2.3 Exceptions to the process standards

The process which represents the entire system (i.e., the Context Level Process) is usually labeled but does not have an identifying number.

Processes which represent major subsystems may not, necessarily, always be namable with a strong-verb and object-noun combination.

5.2.3 Datastores

5.2.3.1 Conceptual standards for datastores

Datastores may neither create nor destroy data.

All data that enters a datastore must eventually leave via some dataflow.

Datastores may not be accessed directly (i.e., through a dataflow) by a terminator (i.e., source or sink).

Datastores may or may not, eventually, become physical files in the implemented system.

Datastores are *temporary* holders of information.

Datastores may not interface directly with terminators without an intervening process.

Data may not flow directly from one datastore to another without an intervening process.

5.2.3.2 Notational standards for datastores

A datastore is represented by a pair of parallel lines.

A datastore must have at least one dataflow into it and one dataflow out of it.

A datastore must have a label.

The datastore label is placed between the two parallel lines.

Datastore labels must be nouns and clearly identify what the datastore contains as a class of objects. That is, instead of labeling a datastore "Target_heading, Target_speed, Target_identification_number, Target_classification_code," use a more concise name such as "Target_Description" and define its composition in the dictionary.

A datastore may appear at any level in the dataflow diagram set at which it is accessed by two or more processes.

The organization of a datastore may not be contained in its label.

5.2.3.3 Exceptions to the datastore standards

If, in the opinion of the software engineer, user, or client, communication would be improved by referring on a dataflow diagram to datastores which exist at a higher level, then reference them by placing their name within parentheses (or other distinguishing symbol) in a position approximating that of sources and sinks.

5.2.4 Sources/Sinks (a.k.a. Terminators)

The terms "source" and "sink" are used interchangeably with the term "terminator." Their use is described below.

5.2.4.1 Conceptual standards for sources/sinks

Sources/sinks are net suppliers or recipients of data.

Sources/sinks lie *outside* the scope of our inquiry—the means by which they derive or obtain data or what is done with the data is not part of the analysis.

Communications of information between or among sources/sinks is not shown on the dataflow diagram, since they are not part of the analysis.

Sources/sinks may not communicate information directly to a datastore without using an intermediate process.

5.2.4.2 Notational standards for sources/sinks

Sources/sinks of information are represented by using labeled squares or rectangles or by placing the name of the source/sink within a pair of parentheses, square brackets, or other distinguishable symbols.

Source/sink labels must be nouns or pronouns.

Sources/sinks appear only at the context level of the dataflow diagram.

5.2.4.3 Exceptions to the notational standards for sources/sinks

Sources/sinks may be referred to at lower levels in the dataflow diagram if communications or clarity would be improved.

5.2.4.4 Systems users, operators, and terminators

One of the more frequently asked questions is, "How should the user of the system be treated—as part of the system or as a terminator?" The easiest and perhaps the most sensible approach is to treat the user as a terminator. That is, this person sends the system requests or data updates and the system sends the user responses. If we choose to include the user as a process, we will, eventually, have to document (via a DFD) just how the user processes information. Given the experiences of Freud, Jung, and Skinner, we are better off dealing with the user as a terminator.

5.3 DEPICTING REAL-TIME PROPERTIES WITH DATAFLOW DIAGRAMS

Today's software systems involve more real-time conditions and responses than ever before. There has been a tendency to incorporate these state-oriented properties into dataflow diagrams via a revised notation. One of the most common revisions of the notation depicts two types of flow on the DFD. The first type is dataflow, the second is control flow. Control flow may be distinguished from dataflow by means of dotted or dashed lines. This has caused a considerable amount of confusion because we view the system in two ways at once and our ability to comprehend what we are seeing can be overloaded. Abuses occur primarily due to the use of this modified dataflow diagram as a flowchart. Remember, flowcharts make control flow explicit while only implying dataflow. Bringing the two views together has resulted mostly in the creation of strong, flowchart-oriented diagrams with the DFD label.

The strongest argument against the incorporation of control aspects into the dataflow diagram is partitioning. The DFD represents the best way of identifying and "clustering" processes based on the information processed. This aspect is more valuable than most authors realize. It is essential that the way in which we "see" the system in order to understand it be divorced, as much as practicable, from the way in which it will be implemented. The incorporation of notation that enables us to view the data-oriented partitioning and control aspects on the same diagram runs the risk of having the dataflow diagram represent an alternate view of the system's modules.

Modules describe *how* the processes will be accomplished. All we are trying to do at this point is identify *what* the processes are. Anything that causes us to consider modules is premature design and implementation. It occurs at the point in time when we are least prepared to make such decisions. Experience on projects where this has been tried has borne this out. This is why we have chosen to keep the dataflow diagram notation, in whatever form that is convenient for you, free of control aspects. The control issues are addressed using Event Diagrams and the Dictionary is used to keep the two synchronized. The issue comes down to either of two tradeoffs. One choice is graphically depicting control and dataflow together at the expense of system partitioning. The other is to depict them separately. This enhances partitioning at the expense of involving an increased effort to keep the two views synchronized. In this text, we have voted for the second choice. Part of the basis for this is that the system will outlive those who develop it.

5.4 CREATING DATAFLOW DIAGRAMS

There are few hard-and-fast rules regarding the use of dataflow diagrams. Unlike control graphs or state diagrams, there is no equivalent to automata theory from which certain desirable properties of a dataflow diagram can be proven to be present or absent. Many firms and individuals have modified the notation and the guidelines to suit specific needs. What is presented here is a starting point from which the reader can develop a tailored style to meet his or her own needs.

5.4.1 A Procedure for Creating DFDs

The basic approach used to develop a DFD is very similar to that used in the IPO (Input-Process-Output) method. That is, we identify the inputs and outputs of the system and relate them to one or more processes performed by it. We will describe the definition process and demonstrate it for a simple physical process— the construction of a taco:

1. Begin by listing all of the inputs and outputs used by or generated by the system or primary process—in this case, to Produce_a_Taco (Figure 5.4.1-1).

INPUT(S)	PROCESS	OUTPUT(S)
– TOMATOES	PRODUCE_A_TACO	TACO
– BEEF OR CHICKEN OR PORK		
– CHEESE		
– SALSA		
– CORN_TORTILLA		
– LETTUCE		
– SPICES		
– TOMATO_SAUCE		
– (OLIVES)		

    Figure 5.4.1-1:   Listing Inputs and Outputs

2. Define the composition of the inputs and outputs via the Data Lifecycle Dictionary (Figure 5.4.1-2).

    LETTUCE = *USUALLY WASHED AND SHREDDED*

    GARNISH = TOMATOES + LETTUCE + CHEESE + (SALSA)

    MEAT_FILLING = [BEEF | PORK | CHICKEN] + SPICES + TOMATO_SAUCE

    SALSA = ["GRINGO" | "STANDARD" | "MACHO"]
                    * STANDARD SALSA REQUIRES A NOTE FROM YOUR DOCTOR *
                    * MACHO SALSA HAS BEEN OUTLAWED BY GENEVA ACCORDS *

    TACO = TORTILLA_SHELL + MEAT_FILLING + GARNISH

    TOMATOES = *THESE ARE DICED OR SLICED*

    TORTILLA_SHELL = * A COOKED CORN TORTILLA   *

    Figure 5.4.1-2:   Forming Data Dictionary Entries

3. Use the output definition(s) to identify the *largest* definable subset of elements which can be identified as being combined to form each output (Figure 5.4.1-3).

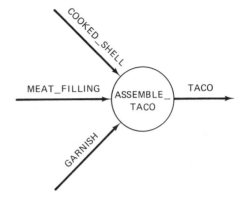

Figure 5.4.1-3: Identifying
Major Dataflow Components

4. Draw a bubble (i.e., process symbol) for each output from the system and give each bubble drawn a label like "Produce _____," or "Prepare _____," indicating that it is responsible for doing the final combination of the dataflows to form its respective output (Figure 5.4.1-4).

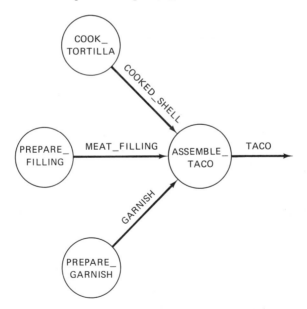

Figure 5.4.1-4: Adding in Candidate Processes

5. Draw each of the dataflows required to form each output (Figure 5.4.1-5).

6. If the dataflows developed in step 5 are the result of the action of a process,

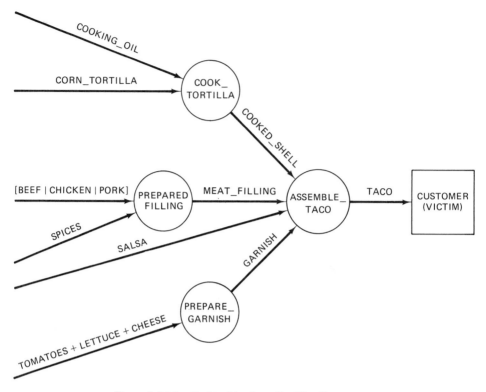

Figure 5.4.1-5: Backtracking from Candidate Processes

place a process symbol (bubble) at the unoccupied end of the dataflows created in step 5. Otherwise, proceed to step 7.

7. Take each output dataflow associated with the blank bubbles created in step 6 and treat each just as you did the system in steps 3 through 6. Name bubbles as best you can before iterating through the loop. Continue this until all inputs to the system have become part of the network or you have run into one or more "dead ends" (i.e., places where further tracing seems impossible). If this happens, begin the link-up of these from the external input side of the system (Figure 5.4.1-6).

A synopsis of the dataflow diagram development process is presented in Figure 5.4.1-7.

There also exists a school of thought that recommends going from the input side to the output side, but this author has found that users of Structured Analysis can create better networks more quickly with the output-to-input strategy. This is particularly true when the data dictionary definitions are done. This output-to-input approach is a lot like the comedy routine where the swami is given the answer to the question and must formulate what the question is. In our case, many different

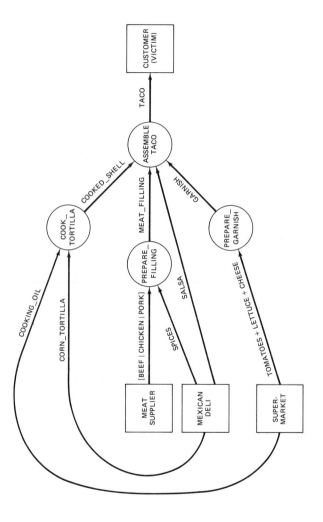

Figure 5.4.1-6: Resulting First-Cut Dataflow Diagram

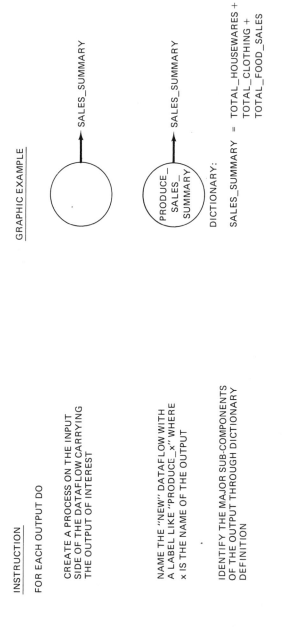

INSTRUCTION

FOR EACH OUTPUT DO

    CREATE A PROCESS ON THE INPUT
    SIDE OF THE DATAFLOW CARRYING
    THE OUTPUT OF INTEREST

    NAME THE "NEW" DATAFLOW WITH
    A LABEL LIKE "PRODUCE_x" WHERE
    x IS THE NAME OF THE OUTPUT

    IDENTIFY THE MAJOR SUB-COMPONENTS
    OF THE OUTPUT THROUGH DICTIONARY
    DEFINITION

    CREATE ONE NEW DATAFLOW FLOWING
    INTO THE PRODUCE_x PROCESS FOR
    EACH SUB-COMPONENT AND LABEL EACH
    NEW DATAFLOW WITH THE NAME OF
    ONE SUB-COMPONENT

ENDDO

The above procedure is repeated for each of the inputs to the PRODUCE_x
process until the source for each data element is identified and accounted for.

Figure 5.4.1-7: Identifying Processes Responsible for Output

questions or explanatory models are possible and acceptable. Hence, we are not trying to emulate some sort of fixed network but are working toward the definition of a model which explains what is observed. This may bear a strong resemblance to what the client is familiar with during the Current Physical Model development.

5.4.2 Complexity

It should be obvious that even relatively uncomplicated systems could result in overly complicated dataflow models. This is likely to happen when the analyst(s) views everything that goes on in a system at a microscopic level of detail. How complex, then, is complex enough? Is there a limit to how much an analyst or client can be expected to comprehend? Fortunately, we have some psychological research to aid us in resolving this question. Its importance should be apparent, since we are going to require that the client give us concurrence on various parts of our analysis. If it is too complex for the client or us to understand, then errors are inevitable.

Some years ago [3] researchers found that the average human can keep track of about 5 to 9 things at once. Hence, 7 plus or minus 2 is seen as bounding the limits of our understanding or comprehension. Thus, it is *advised* that we keep DFDs down to 5 to 9 bubbles. Please note that this is an advisory, *not* a rule or a law! For the purposes of our discussion an advisory is an admonition which, if ignored, *could* result in an unpleasant event. For example, Figure 5.4.2-1 presents a common example of an advisory. Ignoring it does not, necessarily, guarantee that something unpleasant will happen, but it *might*. (My apologies to the various law enforcement agencies, but have you tried doing just 55 mph on the freeway lately?)

```
┌─────────┐
│ SPEED   │
│         │
│   55    │
│         │
│ MILES   │
└─────────┘
```

Figure 5.4.2-1:   An Example of an Advisory

Some systems may have an inherent structure whereby the model(s) we develop may have 12 or 14 bubbles. What then? Merely inspect the model and show it to one or more colleagues. Allow yourself a reasonable amount of time (say, 5 to 10 minutes). If there is a simplification, one of you will spot it. The type of simplification we are talking about here is the collapsing of two or more bubbles to form a single, less detailed one which can be labeled with a verb-noun combination *that makes sense*. If not, then leave it as it is!

Heresy? Not exactly. Remember, what we are attempting to do is to build a model of a system in support of some corporate goal. We are *not* out to create the first and only perfect Structured Analysis. We need to avoid getting overly zealous about this 5-to-9 advisory, since such zeal can be costly. One firm that was using Structured Analysis was so taken by the 7 ± 2 rule that a team of five people met for three and a half hours arguing over how to simplify a 12-bubble diagram. They adjourned without resolving the question. A collective sixteen and a half hours of irreplaceable and costly human resource was lost. This is anything but cost effective. Remember, Structured Analysis is merely a means to an end, not an end in itself.

The appropriate advisory here is: if a diagram has more than 7 ± 2 bubbles, spend 5 minutes reducing it. If it cannot be reduced in that time, then leave it alone.

5.4.3 Leveling and Balancing

As we have seen, with some care, insight, and maybe good fortune, we can define a dataflow diagram which is simple but describes some very complex system. Certainly, such a diagram is of some use, but it is not very informative. A single bubble in a diagram may represent an entire subsystem composed of several hundred simple, interrelated processes. How can we capture that without being overwhelmed by the enormity of it all?

The answer lies in the application of the concept of hierarchy. It is probably the oldest abstract concept humans have employed. Even aborigines have relied on it to explain their relationship to the universe.

The application here is straightforward. The idea is to examine each bubble to see if it could be broken down into simpler bubbles. This is the same approach we used in going from our IPO model to a DFD in the first place (Figure 5.4.3-1). If a bubble can be broken down, we do so in just the way described earlier. We do this

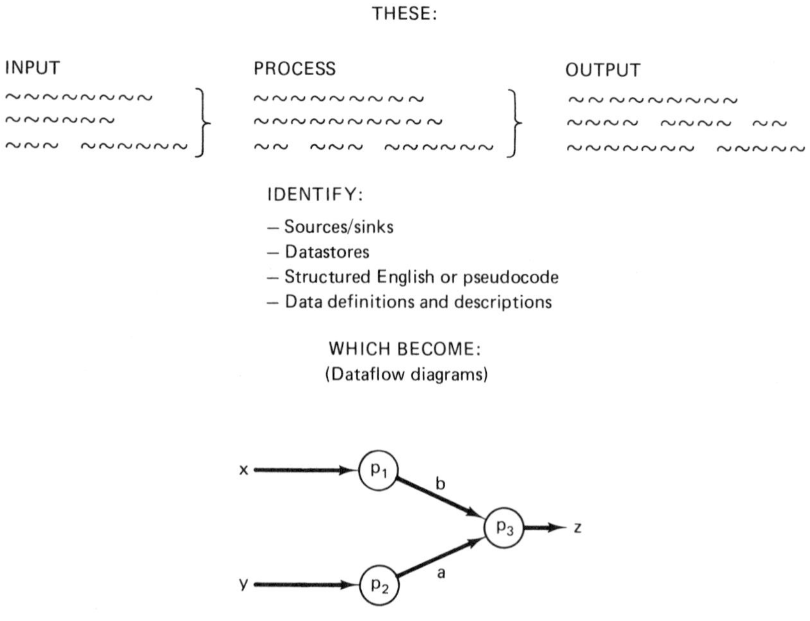

Figure 5.4.3-1:   The Transition from IPO to DFD

for each bubble in the diagram, until every one of them has been decomposed to form part of the next "level" or has been discovered to be a functional primitive. A *functional primitive* is a bubble which cannot be decomposed any further.

The process is depicted in Figure 5.4.3-2. Note that the bubbles which comprise the original or archetype bubble form a family of sorts, since they are "children" or surrogates of the original bubble. These lower-level bubbles themselves can be broken down further (Figure 5.4.3-3) until we have nothing at the lowest levels but functional primitives. Notice that not all bubbles at the top level will decompose the same number of bubbles or processes at the next or subsequent levels. This seems reasonable, since some processes or subsystems at the upper levels are obviously more complicated than others. Remember, primitives (processes that cannot be broken down or decomposed) can occur at any level in the hierarchy. Functional primitives can occur at *any* level of abstraction.

A warning about the concept of primitive processes is in order here. Many who have read the various texts on Structured Analysis or attended some of the seminars have formed the opinion that each process must be decomposed down to the primitive level. This can be hazardous to the health of your project! The problem is that this can result in an overwhelming and unpredictably large number of processes and associated dataflows. This is a project-schedule and resource killer. To put this matter in perspective, of several dozen major efforts the author has consulted on (i.e., first-hand experience) and dozens of projects that colleagues have related, *none* successfully decomposed every process down to the primitive level.

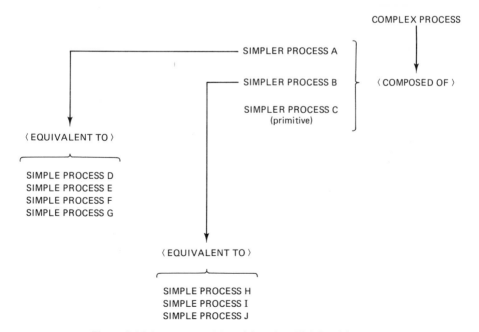

Figure 5.4.3-2:   Decomposition of Complex, High-Level Processes

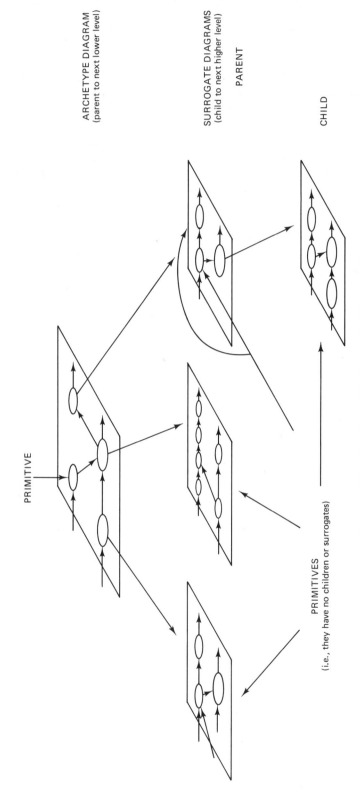

ARCHETYPE DIAGRAM
(parent to next lower level)

SURROGATE DIAGRAMS
(child to next higher level)

PARENT

CHILD

PRIMITIVE

PRIMITIVES
(i.e., they have no children or surrogates)

Figure 5.4.3.3: Reaching the Primitive Level of Decomposition

The use of the primitive concept does constitute a convenient stopping rule for the work in analysis. If it does not work, however, what are some alternative rules? These are described in more detail in Section 5.6. In summary, they amount to a set of strategies for *controlling* the amount of work that must be done in order to create a complete, correct, consistent, and well-partitioned Structured Analysis. They are based on actual experiences and on what works rather than on some ideal approach. The key to them is risk management—that is, to reduce or at least control the amount of risk being taken in order to respond to need and schedule.

For example, the most complicated or complex process is often the one which has the greatest "ripple effect" on the partitioning. "Ripple effect" refers to the phenomenon wherein discoveries made in proceeding from one level to another lower one cause us to change the content of a higher level. The advisories provided in Section 5.5 recommend that the most complex process be addressed first, then the others. In this way, future "ripple" or side effects will be reduced, and, if the project runs short on schedule, people will be rushing through the relatively easy portions of the system rather than the more difficult ones. This reduces errors. However, it runs counter to our natural inclination, which is to put off the most difficult parts of a task until the very end. This inclination guarantees that an unpredictable amount of rework will be required.

Even so, there are many who use Structured Analysis and assume that all bubbles in a system should decompose to the same number of levels. The fallacy of this belief becomes apparent when a relatively simple or primitive process is encountered at a high level in the decomposition.

In order to make it easier to refer to a given level in a diagram by means of a convenient label, each level is assigned a name. The names and their corresponding positions in the hierarchy are presented in Figure 5.4.3-4.

With all of the new bubbles and internal dataflows, it is possible that data may be added or lost in the process. To prevent and/or correct this, we employ another basic principle—conservation. It involves treating each bubble to be decomposed as a system. Just as in the physical world, nothing can be gained or lost in making the transition from representing that system as a single bubble or a network of them. What flowed into and out of the original bubble must flow into and out of its surrogates (Figure 5.4.3-5). Note that the same dataflow names do not have to be used, just the equivalent aggregate of dataflows. Since dataflows are merely pipelines which carry information in one direction or another, the total of all the information carried is the same regardless of the level of detail (Figure 5.4.3-6). Note that the data dictionary definition for each dataflow at one level must in some way involve each of the data items at the next level of detail. Hence, it makes no difference whether or not a particular data item is identified as being optional, iterated, or one of several possible selections (Figure 5.4.3-6).

The point to keep in mind is that we are not dealing with a mathematical abstraction. We are dealing with plumbing! That's right, plumbing. Our concern is with identifying what information is needed by each process, where it will come from, and where the information produced by that process will go. Since data

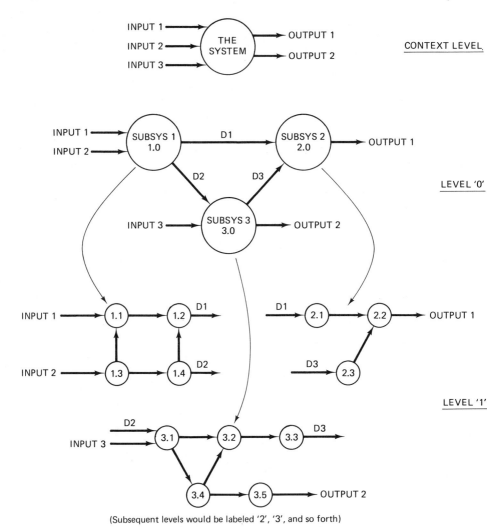

Figure 5.4.3-4: Levels of Dataflow Diagrams and Their Corresponding Numerical Labels

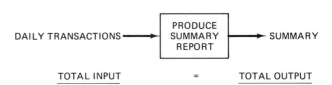

If the balance is maintained, then the summary is directly derivable from the input — nothing more or less.

Figure 5.4.3-5: Conservation of Data in Processes

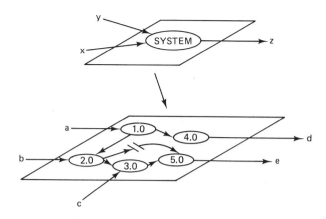

IF x and y are defined in the dictionary to be any combination of a, b, and c

AND

z is defined as any combination of d and e

THEN balance is maintained. Similarly for any process and its surrogates.

For example the following dictionary entries would suffice:

x = (a) + { b }
y = 0 {c} 1000
z = [d | e]

Figure 5.4.3-6:   Conservation of Data Involving Multiple Levels of Dataflows

cannot be spontaneously, mysteriously created, this will entail the identification of flow paths. These flow paths or conduits are referred to as dataflows.

5.4.4 Repartitioning

Although it is recommended in general that leveling be used to decompose systems top-down, top-down is not always the best approach. In fact, the top-down strategy does not work as well as the bottom-up strategy. We are not discussing theory here, but practice. Experience with Structured Analysis (and other methods) has shown that *most* analysis is actually conducted in a bottom-up fashion, with a top-down scheme being used to organize those results. An unscientific study by this author of some recognized "experts" reveals something of a double standard. In lectures, papers, and seminars, these people unanimously state that top-down is *the* way that things are done. So much for the public statements. Privately, these same people admit that they have done most of their own work in a bottom-up fashion. Why the pretense?

What we need to keep in mind is that analysis is an amoral activity—any means to the end (a workable analysis) is acceptable. There, you closet "bottom-uppers"—you can admit your strange practices! Consider this. The information that is obtained during interviews will be obtained at the level of detail at which the person being interviewed chooses to give it to us. If we interview only at the higher levels in the organization, we will get a top-level view, little detail, and questionable accuracy of detailed information. If we ask those at lower levels to give us a higher-level view or "the big picture," they are likely to balk. A clerk probably knows little about how the rest of the store operates, but a great deal about what clerks do. Hence, analysts are usually given information at a low, almost microscopic level of detail. Abstracting this detailed view into a simpler, overall view is the activity we call *repartitioning*.

Upward repartitioning is just decomposition in reverse—synthesis rather than analysis. It involves developing a detailed model based on whatever information has been acquired and examining that model to determine whether or not there are any bubbles or processes which are related by virtue of the nature of the tasks they perform. For example, if we have a group of bubbles with names like "validate_ credit application," "obtain_credit_authorization," "deliver_product," and perhaps some others, they could be lumped together into a single, less detailed bubble with a name like "sell_goods" or "complete_sale." These are not very strong verbs, but they do get the point across. These clusters of bubbles are replaced with their equivalent, and data flows are combined and named in the same way and reconnected. In this way, we can form a level that is one step higher and less detailed than its predecessor (Figure 5.4.4-1).

5.4.5 Top-Down/Bottom-Up: Does It Make a Difference?

The role and importance of Structured Analysis with respect to the design phase, particularly when Structured Design is to be used, can be clearly spelled out here. The concern about whether top-down or bottom-up is a better approach parallels the issue of partitioning and appropriateness of the software design architecture. In the top-down approach one basically imposes one's own view of how the system *ought* to be structured. In the bottom-up approach, to a much greater extent, the system is telling us just what it is structured like. That is, the data elements, their "clusters" or groupings, the processes that use some and create others, and the sources and sinks for this information constitute a description of the system as it is. What dataflow diagrams (supported by the other tools of Structured Analysis) are graphically depicting is just how its various pieces relate to one another.

To many, an important distinction resides in whether things are done top-down or bottom-up. In recent years we have come to accept two assumptions about the top-down concept: (1) it is best and (2) it is a new concept. The second part of the issue can be dealt with rather swiftly. The earliest instance of the application of the top-down approach which has been recorded, translated, and identified occurred about 500 B.C. [4]. A Roman architect was describing for his students how to go about designing a home for a nobleman. He started by considering the site, the position of the sun throughout the year, the surrounding homes, the funds available,

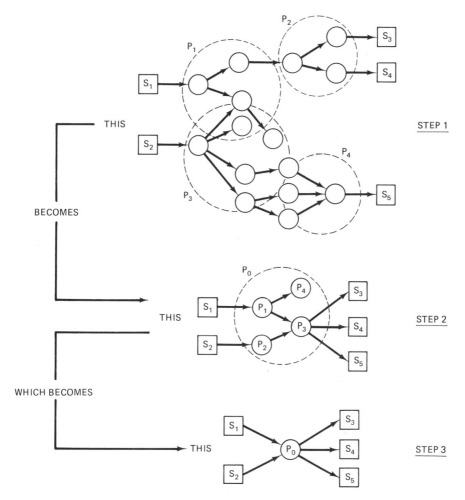

Figure 5.4.4-1:   Upward Repartitioning to Simplify a Complex Dataflow Diagram

and the status of the residents in the community. He then proceeded to the organization and size of the rooms, exterior styles, interior styles, colors, furnishings, and other details. Top-down? Definitely! But the question remains, is top-down best or is bottom-up better?

Perhaps a better question to ask is whether or not either approach can be excluded. In fact, neither can be disregarded, because the data that we obtain will nearly always be at varying levels of detail. Those being interviewed will give us what information they can at the level of detail at which they view it, distorted by their particular perspective. Some information will be minutely detailed; other information will be broad and high-level. Thus, we may have to repartition upward for some of it and decompose some of it in order to achieve a uniform level of detail. In a sense, analysis can be successful only if we are willing to do it like a ping-pong ball—up and down!

5.5 EVALUATING AND IMPROVING DATAFLOW DIAGRAMS

Balancing and leveling dataflow diagrams and developing the corresponding pseudocode, all supported by Lifecycle Dictionary entries, can still result in dataflow diagrams which are of poor quality. This is true even if no software design is planned in conjunction with the development of this problem model. In order to determine whether or not the dataflow diagrams that form such an important part of the problem model are adequate, use the following procedure. It is designed to identify areas of concern where we may wish to focus our attention for productive improvement.

Assuming that an accurate dictionary and pseudocode were developed, the three most common types of problems that dataflow diagrams may exhibit are:

> **Processes which do not transform data.** Developers of dataflow diagrams often label processes with verb-noun combinations which are not transformation oriented. If you will recall, the process is a transformer of data. That is, its purpose is to transform or change input into output. Thus, process names such as "SEND_ALERT" are not acceptable.

> **Dataflows which are actually flags.** Labels of dataflows often reflect a tendency to signal or "flag" an event to a process. Dataflow names like "NO _MATCH" are not dataflows at all but flags.

> **Dataflow diagrams which are actually flowcharts.** The saying, "Old habits are hard to break," appears to be true with respect to many users of dataflow diagrams. In such cases, we find dataflow diagrams which are the result of substituting circles (or other symbols) for the combination of rectangles and diamond shapes one would find in a flowchart.

The above problems often occur in combination, making their correction a somewhat muddled affair. However, the following procedure can be applied to identify what problems exist and where:

1. Make sure that the dataflows have been balanced and the corresponding dictionary is complete.
2. Examine level '0'. Write down a list of the process descriptions.
3. Form the following sentence, "The [fill in the name of the system here] system involves the performance of the following tasks [fill in with the list of processes developed in step 2 above and taken from the level-'0' diagram].
4. If any task or subtask that you can think of that must be performed by this

system is not logically included in at least one of the processes on the list, write it down. Identify which process might be expanded to include it, or create a new process which does include it. This may require some repartitioning of the level-'0' diagram. Revise the dataflows and Lifecycle Dictionary accordingly.

5. Remove any flowchart-sounding names for dataflows by replacing or eliminating them from the dataflow diagram. Revise accordingly.

6. Apply steps 1 through 5 to each of the child diagrams of each of the processes at level '0', substituting the name of the parent process in place of the name of the system in the analysis step (3). Once level '0' is complete, perform these steps on each process and its children and the next lower level as appropriate.

5.6 GUIDELINES FOR STOPPING THE REFINEMENT OF DFDs

Many users of the Structured Analysis method have indicated problems with the use of dataflow diagrams related to the lack of reasonable stopping rules for decomposition. The rule most often applied is to continue decomposing processes until the result is only primitive (or ''atomic'') processes—processes which cannot be broken down any further. This ''stopping rule'' has caused many problems:

Some processes are simple enough that it makes no sense to require breaking them down to the same level of detail as others that are more complex.

Breaking processes down to this level consumes a lot of resources. In the case of simpler processes, it is not apparent that such expenditures are warranted.

The nature of the system may be such that we may not fully understand what is required in terms of people resources and will experience some sort of scheduling problems, or the nature and number of requirements changes from the customer may aggravate schedule difficulties. In such a situation, does it make sense to detail all processes when there are only a few (typically three or less at level '0') which will ''drive'' the system's structure?

The observations above have been made during dozens of project consultations. People using these methods need a more flexible and practical stopping rule than the ''atomic process'' guideline. A guideline is offered below as a means of meeting this need. However, the ''atomic process'' guideline *is* effective *if* there is enough resource in terms of both flowtime and person-hours. Hence, what is presented below should be viewed as a backup approach. It has shown itself to minimize the time required to decompose and refine dataflow diagrams but ensure that what is produced has the maximum quality, given the constraints of stringent schedule and personnel requirements:

1. Develop the level-'0' dataflow diagram complete with Lifecycle Dictionary entries.

2. Examine the individual processes at level '0'. Identify those that are simple enough to be clearly understood (a judgment call on the part of the analysis team) and the one or more that are clearly more complex than the others.

3. For those that are clearly understandable, write pseudocode. If the pseudo-code is greater than 50 to 100 lines long, refine that bubble to a lower level.

4. For those that are clearly more complex than the others, decompose them one more level. Examine the results of this decomposition for each of the ''parent'' processes at a level by proceeding to step 2 above, and work your way through the process again until the results are simple and understandable.

Note that the overall effect of the above approach is to focus our energies on those portions of the system which have historically caused the most difficulty and to minimize the amount of energy (and money) expended on portions which have historically caused the least. In this way, we can control time flow/schedule and costs while maximizing the quality of the results.

5.7 AUTOMATING THE USE OF DFDs

Anyone who has considered the use of DFDs on complex, multileveled systems must agree that the originators of this notation may have never considered how or whether it could be automated. Consider the problems: we are linking circles with arbitrary curves labeled with text that parallels the arbitrary curve—not exactly a problem for a class on Introduction to C Programming.

Vendors who are developing and promoting automated tools to support the use of Structured Analysis and particularly DFDs have made some significant decisions. One is to base their tools on the IBM-PC and compatibles. This is not a bad decision, since such hardware is inexpensive and in widespread use. Another decision is not a good one: it is to assume that the DFD notation is sacrosanct—that it cannot be changed.

Some have adopted a form of DFD [5] which employs vertically oriented rectangles with rounded corners and dataflows which travel vertically or horizontally, making the transition from one to the other with a rounded rather than a sharp corner. So much for cosmetics. Let's look at the basics.

First and foremost, a DFD is a network, and networks can be easily represented by means of matrixlike tables. They do *not* require the use of *some* arbitrary graphic. This is particularly true when one considers that the graphic is meant to show a concentrated representation of the culmination of the analysis effort. It is only a picture and a worthless one at that, unless the dictionary, pseudocode, and other supporting materials are present and consistent with it. Unfortunately, vendors have become really ''hung up'' on this graphic. They fail to note that the DFD is only as good as its supporting material—*not* its graphic quality. Hence, many have let error checking suffer at the expense of pretty pictures. More about this in later chapters.

There are alternate representations of DFDs that would be easier to automate and would maintain the network nature of the method. However, automated versions of the DFD in any form must provide, as a minimum, the following set of features:

- Ability to add, modify, or delete a process, dataflow, datastore, or source at any level at any time. This would include automatic renumbering and other adjustments.
- "Preview" or look ahead as to the impact of a change before it is made. The system would present a menu of which processes, data items, etc., would be affected by the change just entered by the user. At that point, the user could cancel the change or confirm it.
- *Automatic* balancing, leveling, and pseudocode cross-checking as the information is entered. Upon our striking the enter command, the pseudocode (for example) would be checked against the dictionary, the process it represents, and the dataflows flowing into and out of the process to ensure that everything is correct and consistent. For example, all data items in the pseudocode are defined in the dictionary *and* vice versa.
- User-selectable or convenient limit (for example, 32 characters) for the names of dataflows, datastores, processes, sources/sinks. Most tools now have a limit of 6 or 8 characters in a name or only recognize the first eight characters. This leads to unintelligible abbreviations.
- User-selectable or convenient limit (for example, no limit) for the number of processes that can be present on a DFD. Some tools currently limit the user to 7 ± 2 processes. This is ridiculous! On many occasions 10 or 12 or even 14 processes have seemed perfectly reasonable and understandable to all concerned. This 7 ± 2 limit is another example of religious fervor gone wild.
- Output of *true* report/document quality. That is, the *entire* analysis package should be generated by the system in a presentation format. This format includes but is not limited to a table of contents, automatic section and page numbering, and an automatically generated index for all data items, processes, and other elements of the analysis package. Customization by the user on a project-by-project basis is also required.
- The tool must support Event, Process, and Information Modeling *and* automatically perform cross-checks among them.
- The tool must *automatically* generate a first-cut structure chart.

The key thing to keep in mind about any automated tool to support Structured Analysis is that Structured Analysis is a means to an end—*not an end in itself.*

REFERENCES

1. HIPO—A Design Aid and Documentation Technique, IBM Corp., Manual No. GC20-1851. White Plains, N.Y.: IBM Data Processing Div., 1974.

2. H. Katzan, Jr., *Systems Design and Documentation: An Introduction to the HIPO Method.* New York: Van Nostrand-Reinhold, 1976.

3. G. A. Miller, "The Magical Number Seven, Plus or Minus Two: Some Limits on Our Capacity for Processing Information," *Psychological Review,* Vol. 63 (1956), pp. 81–97.

4. W. R. Spillers, ed., *Basic Questions of Design Theory.* New York: American Elsevier, 1974.

5. C. Gane and T. Sarson, "Structured Systems Analysis: Tools & Techniques," Improved Systems Technologies Inc., New York, July 1977 (ISBN 0-931096-00-7).

6. S. Hori, CAM-I Long Range Planning Final Report for 1972, Illinois Institute of Technology Research Institute, Chicago, December 1972.

7. D. T. Ross and J. W. Brackett, "An Approach to Structured Analysis," *Computer Decisions,* Vol. 8, No. 9 (September 1976), pp. 40–44.

8. "An Introduction to SADT Structured Analysis and Design Technique," Sof-Tech Inc., Document No. 9022f-J78R, Waltham, Mass., November 1976.

9. "SAMM (Systematic Activity Modeling Method) Primer," Boeing Computer Services Co. Document No. BCS-10167, Seattle, October 1978.

CHAPTER 6

Policy Statement Methods

Thus far, we have described techniques for defining the information and the relationships inherent in that information's use in the enterprise, and we have discussed one means of describing some of the operating rules within the organization. In this chapter we will expand our notion of system operating rules into more formal, process-oriented descriptions of how policies are implemented. We will examine two basic forms: one is textual (pseudocode, structured English), the other graphic (decision tables, decision trees). Each has its strengths and situations where it is most effective. None is effective for *all* situations.

6.1 STRUCTURED ENGLISH AND PSEUDOCODE

The profession that perhaps comes closest to the role that software development plays in most corporations is accounting. The software developer classically has to learn how a particular part of the operation works in order to automate all or part of it. Anyone who has been through this sort of activity has probably experienced difficulty in communicating with the organization being helped. There are a lot of ways that the people being studied respond. Their responses range from feelings of rejection and hostility to an attitude of arrogance and include just about everything else in between.

   The arrogance often emerges because the systems analyst does not understand their part of the company as well as they do. Clients have related many incidents in which systems analysts have essentially tried to tell them, the client, how to run their own business. This classic communications problem stems from the fact that the client and the systems analyst tend to "see" the world through different eyes. They are each sensitive to different things, and they explain the world via their own personal and professional models.

This communications difficulty is aggravated by the fact that company policy must be reflected in the analysis. Why aggravated? Because many of the activities which occur within the organization are not fully understood by the analyst *or* the client. The client personnel know something about how a process works, but the analyst needs to know *all* about how it works, what it needs, and where that information comes from. Surprisingly, this occurs just as often in situations where the analyst and the client are both employees of the same corporation. One would presume that these people would have some common knowledge base and the communications difficulties would be lessened, but such is not the case. These problems are increased by the rate of change in the business itself as well as our everyday lives [1].

So, into this maelstrom comes the systems analyst, armed with dataflow diagrams and a textbook, only to find that a considerable amount of information about this "system" has not been written down. The analyst may know a great deal about computing hardware and software but very little about a particular type of business. This is even true if a similar system was studied and built for another firm. Terms differ, priorities are different, accounting systems are specialized. Similarly, the businesspersons associated with this project know the enterprise in an intuitive as well as a conscious sense. Their response to many questions will be, "There is no reason for it, it is just our policy!" Hence, both sets of participants in the project view the other as both superior and inferior as well as somewhat suspect.

If we add into this situation the latitudes of interpretation possible with a natural language such as English, we have a situation tailor-made for miscommunication.

Structured English and pseudocode provide us with a means of avoiding miscommunication. Each is well suited to a specific type of situation. Together they provide us with the flexibility to meet a wide spectrum of problems. Structured English meets our need to document client policy in an Englishlike manner. Some clients prefer this style. Other clients prefer to have policy descriptions documented using a more codelike format, pseudocode. This latter approach provides us with a means of avoiding the need to convert or refine the structured English to pseudocode. Overall, the use of pseudocode is both a labor saver and the more accurate approach of the two.

6.2 STRUCTURED ENGLISH

The first use of the term "structured English" may have been by Caine and Gordon in 1975 [2]. They described it as a means of solving the communications problems that existed between business clients and software developers. Many of us have had experiences wherein we delivered a quality system to the client but the client was not satisfied with it. When we investigated, we found that an elegant solution had been provided but for the wrong problem. That is, the problem we solved so well was not the problem the client needed to have solved.

Caine and Gordon found that a way to avoid this was to employ a series of successively refined and detailed statements of what the client had to have the system do. This increased communication. This process of refinement would continue until there was enough detail that the code would be obvious. They did not offer a system architecture concept similar to that of Structured Design but they did offer designers and analysts this strategy. The key aspect of this approach is that complex policies and procedures and the jargon of the business being studied are simplified. They are simplified sufficiently to allow someone (the analyst) to understand a business which was formerly alien to him or her. This simplification process also requires that the process be partitioned. This reduces the problem of explanation to one of describing one part or partition only. Structured Analysis provides a means for partitioning via the identification of processes and transformations.

Structured English meets the needs of Structured Analysis to describe the procedure being carried out within each transformation in order to accomplish the required change or transformation of data. What we are actually trying to capture with structured English in Structured Design is what the client's policy is. For example, how the client wants certain computations done, what the rules are regarding the validation of input data, what the limits are with respect to overdue payment, how to compute interest and under what circumstances it should be charged—all are examples of client policy. It may vary from client to client.

6.3 RELATIONSHIP OF STRUCTURED ENGLISH TO THE OTHER TOOLS OF STRUCTURED ANALYSIS

Structured English enables us to capture customer policy in clear, concise terms. Although many different styles may be employed to do this, they all have certain things in common with the rest of Structured Analysis and Structured Design. Figure 6.3-1 graphically depicts the relationship of structured English to the rest of the tools. A few basic rules are:

Each data item mentioned in the structured English is defined in the data dictionary.

Structured English is required for all processes which cannot be further decomposed. Often, however, we do not have sufficient time, patience, or personnel to decompose every process down to its functional primitive level of detail. As a result, if we follow this "minimum" rule, we may end up having to go to the design phase with an analysis package which lacks structured English for many of the processes. Hence, it is recommended that structured English be completed for processes at a given level before

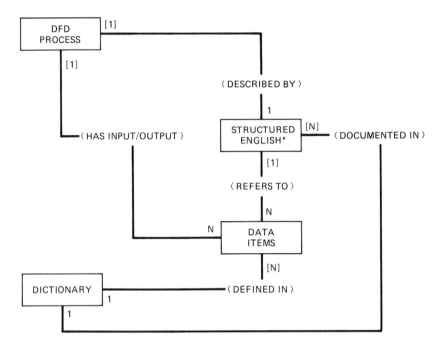

Figure 6.3-1: Relationship between Structured English and Other Structured Analysis Tools

decomposing to a lower one. In this way we utilize iterative refinement to best advantage and insure ourselves against some of the problems associated with schedule changes and slides.

6.4 DIFFERENCES BETWEEN STRUCTURED ENGLISH AND PSEUDOCODE

The main purpose of structured English is to express the client's policy in a disciplined manner which the client can validate. Unlike pseudocode, it is not intended as some sort of simplified statement of a computer program which could be coded from. Although no standard exists for either means of describing a process, there are some important differences between them.

Structured English emphasizes how the process works in an abstract sense. That is, error traps and responses are downplayed, particularly minor errors, while the mainline process is clearly the focus. The emphasis is on what a process like BILL_CUSTOMER does rather than on how it is done (Figure 6.4-1). Note that although we are computing and replacing the values different variables have, we are not told how or whether files are opened or data is read from some device, what the

STRUCTURED ENGLISH	PSEUDOCODE
Policy oriented	Implementation oriented
Statements use terms familiar to client (business language)	Resembles programming language
Objective is high level of communication	Objective is to support programming language realization

Figure 6.4-1:   Differences between Structured English and Pseudocode

computation and roundoff rules are, as well as other important aspects of this process, including error processing and error messages. These are more the province of the pseudocode developed during design. It is employed in such a way as to detail how the process may be realized on some sort of abstract computer. That is, it is temporarily assumed that there is programming language which has the same characteristics as the pseudocode. Although the use of an actual programming language, such as ADA* [3], has been proposed as an effective means of employing pseudocode, there are some serious problems with this. These include our being in a stage of system development where we are attempting to attain an accurate and concise understanding of just what the problem is. But more fundamental problems make the use of an actual programming language an imprudent and ineffective choice for describing a generalized set of rules about how a process is accomplished.

There has been considerable study of how the human brain works. One area of concern here is the means by which creative or new ideas are generated and externalized. In this regard, the left- and right-brained model [6] has become popular. It states that the human brain has delegated different roles to each of its halves. The left half is devoted primarily to the use of logic, while the right half is responsible for intuition. The right side has been given credit for the "insightful flashes" one sometimes gets when looking at a troublesome problem, explaining it to someone, and suddenly seeing how it may be solved. The use of a programming language with its inherent syntax and semantics relies primarily on the left side of the brain. However, one of the most important reasons for using pseudocode in the first place is to identify possible errors in the information given us by the client. Hence, the use of a programming language such as ADA would tend to proceduralize what we have been told, but it would do so in a way that severely restricts our intuitive skills at determining whether or not what we have been told is reasonable. Although the client's description may result in an *implied* model of how a problem may be solved, this does not mean that we should identify solutions before we have even explored alternative solution models. If we use a programming language with its syntactic and semantic restrictions, that is just what will happen. When we employ an expression mechanism such as a programming language, we utilize the sequen-

*ADA is a trademark of the U.S. Department of Defense.

tial processing part of the brain, *not* the intuitive or creative part. Are we ready to forego other possibilities at this point?

6.5 STYLES OF STRUCTURED ENGLISH

There are nearly as many styles of structured English as there are analysts using it. Some people prefer a style which is pseudocodelike, while others prefer an Englishlike style. We will present and explain several different styles, letting the reader select the one(s) best suited to a particular analysis situation:

> Pseudocode style
> Outline style
> Prose style

No particular style should be thought of as being the best or worst. Rather, the deciding factor should be its relative ability to communicate with the client. Some clients are "turned off" by anything approaching code, so the more proselike style would be better than pseudocode. We all have our favorites—but again, the objective choice is an important one.

6.5.1 Pseudocode Style

This is probably the most popular style of structured English. As the name implies, it takes a form which approximates code. This style employs a simple set of constructs (IF-THEN-ELSE, SEQUENCE, and WHILE-DO/DO-UNTIL). Perhaps the reason for its popularity with analysts is that most analysts are former programmers. Hence, they exhibit latent programmer tendencies! Some clients are put off by this style because

> They are not programmers and as such are not used to having to clearly and concisely state policy.
> They do not see their business procedures as a set of sequential processes but rather as a set of continuous activities not expressible in such simple terms.

The advantages to the software engineering organization include

> Simplicity and conciseness.
> Ease with which we can develop the pseudocode in the Structured Design phase.
> Relatively rapid identification of logical errors and inconsistencies in customer policies.

PROCESS: Inform Prospective Clients

WHEN a request_for_service is received from a client:

First, check to see if they have an account_#

    IF the account_# is classed "INACTIVE" then
    reactivate account via procedure #438

    IF the account_# is classed "ACTIVE" then
    determine type of service desired:

        WHEN service = "purchase"
        read them service-agreement

        WHEN service = "lease"
        read them synopsis of lease agreement

        WHEN service = "subcontract"
        arrange for an appointment

    IF the account_# cannot be found then
    open a new account_# via procedure #6A

Figure 6.5.1-1:   Example of the Use of the Pseudocode Style of Structured English

Most users of Structured Analysis have found the advantages far outweigh the disadvantages. An example of the pseudocode style is presented in Figure 6.5.1-1.

6.5.2 Prose Style

This style of structured English is not very popular. Ideally, it is a modified form of english in which there are no adjectives or adverbs. An example of this form of Structured English is presented in Figure 6.5.1-2. It uses the logic from Figure 6.5.1-1 for comparison.

PROCESS: Inform Prospective Clients

When a request_for_service is received from a client, check to see if the client has an account_#. If the account_# is classed ''INACTIVE'' then reactivate account via procedure #438. If the account_# is classed ''ACTIVE'' then determine type of service desired. When the service desired is ''purchase,'' read the service agreement. When service desired is ''lease,'' read the synopsis of lease agreement. When service is ''subcontract,'' arrange for an appointment. If the account_# cannot be found, open a new account, assign a new account_# via procedure #6A.

Figure 6.5.1-2:   Example of Prose Style Structured English

6.6 PSEUDOCODE STANDARDS

Pseudocode is an alternate means of describing policy. At present there are no published standards for use with pseudocode. The subsections which follow represent a consistent set of guidelines or standards to fill this need.

6.6.1 Conceptual Standards for Pseudocode

Pseudocode describes policy but is intended to simplify and clarify policy

Policy may take any of several forms, including mathematical equations, textual statements, and codelike procedural descriptions

During the analysis phase, pseudocode describes *what* the processes do without describing *how*. For example, stating that the name of a client will be found by the system given the client's code number describes what, but to also relate that an index sequential search would be performed for some inquiry would be describing how.

During the physical design and implementation phases, pseudocode will incorporate how the processes are carried out.

It is recommended that pseudocode be developed for each process in the dataflow diagrams that are part of the analysis. As a minimum, pseudocode should be created for the primitive-level processes in the analysis model.

During the design phase, pseudocode should be developed for each module.

The variables or data items which are used in the pseudocode must be defined in the Lifecycle Dictionary.

6.6.2 Notational Standards for Pseudocode

Pseudocode is a codelike process-statement language for which no industrywide standard currently exists. Its basic content is the three fundamental constructs:

```
IF-THEN-ELSE
DO-WHILE AND UNTIL-DO
SEQUENCE
```

The data element names incorporated into the pseudocode must be the names of data items that are defined in the Lifecycle Dictionary.

The pseudocode must be indented to indicate nesting levels or control gates necessary to reach a particular section of it.

The pseudocode for a given process (or module during the design phase) should not exceed approximately 50 lines (i.e., one typed page), single-column format.

The pseudocode for a given module in design must include a CALL statement for each and every module that is called by the module being described. All required logic for the module to properly perform its tasks must be included. The CALL statement(s) must include enumeration of the data that is sent from the calling module to the called module and the data that is returned from the called module to the calling module as follows:

        CALL MODULE X (I1,I2, . . . ,In/R1,R2, . . . ,Rm)

where I1 through In are data sent from the calling module to module X and R1 through Rm are data returned from module X to the calling modules.

An example of the use of pseudocode is presented in Figure 6.6.2-1.

```
IF APPLICANT_AGE LESS THAN 40 YEARS
    THEN
        IF APPLICANT SMOKES
            THEN
                USE RATE #1 INCREASED BY PENALTY_PERCENTAGE
            ELSE
                USE RATE #2
        ENDIF
    ELSE
        IF APPLICANT SMOKES
            THEN
                USE RATE #2 INCREASED BY PENALTY_PERCENTAGE
            ELSE
                USE RATE #1 REDUCED BY INCENTIVE_PERCENTAGE
        ENDIF
ENDIF
```

Figure 6.6.2-1:   Example of the Use of Pseudocode

6.6.3 Exceptions to the Notational Standards
for Pseudocode

Pseudocode may be developed for processes defined in the analysis phase which are primitives. That is, they cannot be broken down further. If the number of levels the system is decomposed into is limited and/or the likelihood of high-level changes low, then only the lowest-level processes need be described by pseudocode.

6.7 OTHER POLICY STATEMENT METHODS

Some policies and operating rules are much too complicated to comprehend using textual methods. Such complex situations can effectively be addressed by tabular methods which display all possible situations and outcomes. These approaches enable us to determine whether or not the proposed software system will be able to respond in an acceptable manner to any and all sets of conditions encountered.

It is often the case that, since a condition occurs only occasionally, the client may not have considered it worth mentioning. In many cases the client may not even be aware that such a condition can occur. A perfect example occurred during the development of a retirement system for a state. State employees could call in to an operator, who would input data into the system. It would display the inquirer's current benefits, what they would be at earliest point of retirement, and what they would be at age 65. Owing to a considerable amount of lobbying and a natural desire by many state legislators to ensure their getting excellent retirement benefits after only a few terms, the retirement law became a morass of if's and whereas's. The situation was so complex that representatives from the state attorney general's office were unclear about which combinations of city, county, and state employment were qualified under the state's retirement law and which were not. The use of a tabular approach to resolving what employees and types of service were covered did not occur until the design of the system was well under way. The table was so large that, using one-inch quadrille paper, it covered an area five feet high by two feet wide.

Long before the results were complete, it was determined that a large number of state employees who were contributing to the retirement system were actually *logically precluded from receiving benefits*. Such an employee could contribute for many years, call in occasionally and find out what benefits would be received, and feel that nothing was out of the ordinary. At the time that benefits were to be paid, however, the qualification algorithm would exclude payment of benefits. As you might expect, this revelation was not enthusiastically received. In fact, the official position of the attorney general's representative was that "that cannot be right!" Fortunately, cooler heads prevailed, the problem was acknowledged, and several modifications of policy were put into place to correct it.

All of this occurred *after* the retirement policy had been transformed from "legalese" into structured English. This does not mean that structured English is useless. What it does indicate is that more than one means of stating policy is advisable. This is particularly true if policy is complex or involves many different organizations. As we have seen in earlier discussions (see Chapter 4, "Entity-Relationship Diagrams"), policy is a difficult thing to capture. What is worse, policies may vary from one organization within the company to another. Automation can aggravate this situation by making it apparent rather than keeping it hidden.

6.8 SOME ALTERNATE MEANS OF DESCRIBING POLICY

Structured English provides a means of stating policy which is more concise and precise than everyday conversation. However, it can become cumbersome when a policy or procedure is complicated. This is particularly true where the policy involves several actions related to several sets of conditions. Most organizational policy is a result of a combination of industry practice and response to unanticipated conditions. As a result, there can and have been many instances of manual systems which were cumbersome but worked. When they were replaced with automated ones, failures occurred frequently. Hence, some means of graphically, logically displaying policy is needed which enables the analyst to detect "holes" or logical errors. For example, an airborne radar tracking system was designed, built, and tested before it was accidentally discovered (via a system "crash") that it was not designed to track flying objects which could stop in midair and reverse direction (such as helicopters). The system had a logical "black hole." Tools that can reveal such problems before they become expensive embarrassments include:

Decision tables
Decision trees

Both of these tools provide a tailored way of depicting decision- or policy-oriented information. They are described in the sections which follow.

6.9 DECISION TABLES

The use of decision tables dates back at least to the initial attempts at switching systems by the then fledgling Bell Telephone Company, and it may have started with them. The decision table may take many forms [4, 5], each with its own special area of application. In its basic form, the decision table is composed of a condition section and an action section.

The condition section lists all of the possible conditions or events which may occur singly or in combination. The main thing is to get an accurate and complete list of *all* possible conditions. Some conditions may preclude or exclude others. For example, if the system is for a hospital, we may have conditions such as

1. Covered by private insurance
2. On welfare
3. Age 65 or over
4. Second party responsible
5. Prepaid
6. Donated service

as well as others. Note that if certain conditions are true, then certain others no longer have any applicability.

Actions are responses or policy statements related to conditions or sets of conditions. In preparing the action section of the decision table, the emphasis again is on completeness and accuracy. All possible actions should be listed. No system response will occur that is not listed.

The obvious value of decision tables is that they help the analyst (and designer) sort out complex and obscure policies to find out whether they lack coherence and where. A not-so-obvious side effect is the message that decision tables may contain for the client. Often, they will reveal that policy has not been refined and that a great deal of on-the-spot decision making—policy making—is going on without the knowledge of corporate management. Another side effect is that of bringing both client and analyst "down to earth" by putting things in perspective.

The value of the use of decision tables was exemplified recently in a situation involving a power company. The company wanted to reduce the manual labor required to generate letters to clients who had their electric service discontinued when they moved out but did not pay the final bill. These people were called "nonzero shutoffs." People in the service area of this firm had learned that they could have the electricity shut off when they moved from one location to another without paying the final bill. However, the power company had a policy that outstanding bills must be paid before new service could be started. Since the company's area of service was quite large, it was likely that these "nonzero shutoffs" would eventually apply for service with this same company. Because the company had a monopoly on such a wide area, the public utilities authority had created a virtual labyrinth of rules regarding what action(s) could and could not be taken by the company, once such a person was found.

About two dozen people worked these nonzero-shutoff cases, knew the rules, and used forms and coded numbers to indicate what type of letter should be sent and what other action, if any, should be taken. Discussions regarding these policies were filled with ifs, ands, and exceptions to previously stated rules. Out of a desire to resolve this logical nightmare came the suggestion that a decision table be drafted which would spell everything out. The client could not understand why, things seemed simple enough. However, when one was drawn up, it was based on the conditions the client said were possible. We used quadrill paper which was ruled into one inch squares. Based on what we were told by cognizant power-company employees, we would have needed a length of paper sufficient to go from their eighth floor office where we were working down to street level, across the street, and into a neighborhood pub! Appraised of this, the client began to admit that, in fact, there really were some conditions which logically "locked out" others. The final result was still over sixteen feet long but a far cry from the original prospect. This experience highlights two important properties of decision tables:

They can be used to identify logical inconsistencies contained in policy. The power company staff was unaware of it but there were several instances which could occur for which policy was not stated. So how did this operation continue? That's easy, people simply made a personal decision (ad hoc policy-making). That enabled the company to continue to recover potentially lost revenue but it also meant that management had no way of knowing with certainty that what was being done was in compliance with applicable statutes.

They put "complex" policy into perspective. What we were constantly being told was that this rule or that rule definitely applied. But this opinion varied from one client contact to another. Upper level management felt that the whole thing was quite simple. The one hundred foot or so paper length brought both parties to their senses and it was "discovered" that not all rules always applied. In retrospect, a certain amount of job protection may have been present.

Let us use a simpler example to demonstrate the use of the decision table. We shall look at an insurance company's rate-making policy. Figure 6.9-1 displays one

CONDITIONS

MALE	X	X	X	X					X	X		
FEMALE					X	X	X	X			X	X
UNDER 40 YRS OLD	X	X			X	X			X		X	
40 YRS OLD OR MORE			X	X			X	X		X		X
SMOKES	X		X		X		X					
SMOKED BUT QUIT									X	X	X	X

ACTIONS

USE RATE 1	X	X			X	X						
USE RATE 2			X	X			X	X				
INCR. RATE 10%	X		X									
REDUCE RATE 10%						X		X				

Note: We are not told what to do about former smokers.

Figure 6.9-1:   One Form of the Decision Table

form of the decision table. The five conditions listed are presented in what is referred to as the "Condition Stub." The actions have been presented in the "Action Stub." Actions are actually policy(s) associated with each condition or combination of conditions. Note that in the general case, there may be a condition or set(s) of conditions which are logically possible but for which no policy has been provided. Though they may have been rare occurrences in the experience of the hospital, a software system which encounters these circumstances and has not been programmed to respond rationally to them will fail.

6.10 DECISION TREES

Decision trees are an alternative way to represent the same set of information contained in the decision table. They differ primarily in that, whereas the decision table is a matrix representation of the information, the decision tree is a network-oriented one. The same set of conditions and actions presented in Figure 6.9-1 are

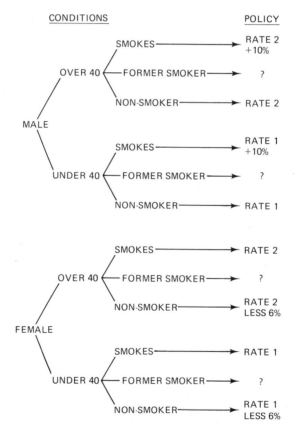

Figure 6.10-1:   Example of a Decision Tree Based on Figure 6.9-1

presented in the form of a decision tree in Figure 6.10-1. The use of either form is a matter of personal preference, although the decision table appears to be more compact. Recently, research has been published which shows that under certain conditions, the decision tree is superior to either the decision table or structured English.

REFERENCES

1. Alvin Toffler, *Future Shock*. New York: Random House, 1971.
2. S. H. Caine and E. K. Gordon, "PDL—A Tool for Software Design," *Proceedings of the 1975 National Computer Conference*, Vol. 44. Montvale, N.J.: AFIPS Press, 1975, pp. 271–76.
3. M. Montalbano, *Decision Tables*. Palo Alto, Calif.: Science Research Associates, 1974.
4. H. McDaniel, *An Introduction to Decision Logic Tables*. New York: Petrocelli Charter, 1978.
5. Betty Edwards, *Drawing on the Right Side of the Brain*. Los Angeles: Jeremy P. Tarcher, Inc., 1979.
S1. Extracted from the seminar, "Structured Analysis for Real-Time Systems," by Software Consultants International, Ltd.; Kent, Washington; Copyright 1988. Reprinted by permission.

CHAPTER 7

Event Diagrams

In our earlier discussions of dataflow diagrams, we treated the network of interacting processes as though each one were "on." That is, we ignored the possible situations in which one or more of these transformations would not be appropriate to the type of processing being conducted. We noted that, in a sense, there was a mapping of processes that were appropriate to given situations; the mapping would change depending on the situation. This mapping controlled when these processes were on and others were off. The mechanism by which this on and off status is changed and maintained is that of *change of state*.

In this context we will define a *state* to be a mode of operation wherein certain processes or activities are appropriate and others are not. The mechanism by which the system changes states will be referred to as an *event*. Hence, in order to understand multistate systems (e.g., real-time, interactive), all we need to model is the relationships between states and events in addition to the population of those states in terms of processes. This combination will enable us to describe what events cause the system to change states, what the states are, and what processes or activities are appropriate for various states. This is a somewhat different view than has been used recently and in the mid-1970s to describe such systems. In those periods an attempt was made to incorporate the mechanism, the control information, the input data, and the output data into a single diagram [1]. Although that may seem to be a powerful graphical approach, in fact few people could fathom the diagram well enough to feel comfortable about declaring it to be correct or not.

In this chapter we will describe a simple means of describing multistate systems. It is graphically simple and, more importantly, is an integral part of our three-dimensional view of systems. We will also demonstrate how this type of diagram can be derived from dataflow diagrams and made consistent with the E-R diagrams. The discussion includes a step-by-step procedure for deriving these diagrams.

7.1 EVENT MODELS

Event Models are intended as a means of describing real-time systems. They provide us with a means of describing a dimension of software systems which is usually ignored. This is particularly true if one requires that the Event Model have *some* relationship to the Process Model.

Event Models describe three subproperties of real-time systems:

States. The modes of behavior the system can operate in.

Events. The occurrences that cause the system to change its mode of behavior.

Concurrence rules. The policies that will be invoked regarding which states may coexist and the rules about how states may be reached.

All of the above are described in the Lifecycle Dictionary. During the Analysis and Logical Design phases, the Events are described with respect to composition. During the Physical Design phase, the Events may become interrupts, with their bit patterns being described in the Physical Design and Implementation phases.

7.1.1 System States

Many systems have several different modes of behavior. For example, in using personal computer applications software, we have experienced the phenomenon of inputting a perfectly valid command to the system and getting an error message in reply. After a moment's thought, we discover that our error was not that we misentered the command but that the command was not valid for the mode the system was in. Entering the logon command when we are already logged in qualifies as one of these miscues. Note that the acceptability of the data that we are inputting does not rely solely on the data itself (i.e., its values) but on some other factor that has nothing to do with the data per se. This other factor is what we shall refer to as the *state* the system is in. Perhaps this is how multistate systems should be defined:

Multistate systems are such that the acceptability of input data is a function of the data itself *and* an internally maintained map which changes due to events external to the system.

7.1.2 Events

Events are occurrences which influence the way in which the system functions. Examples include I/O interrupts, timeouts on physical devices, warning or error inputs, the exceeding of a threshold by a suspected target, and the verification that a target track is complete. Each can cause the system to change its behavior in a fundamental way. These events are considered by us to be external to the system, because that is in the very nature of such systems. They respond to their environment. Sometimes they respond in unexpected and unpleasant ways.

7.1.3 Concurrence Rules

Some systems do not just "jump" from one state to another because of external actions but have more than one state in existence at one time. The mechanism by which this is accomplished is not of concern here. Our concern is that the system we are modeling must be able to support more than one state at a time. This can be represented as a decision-table-like mapping of compatible and incompatible states.

7.2 EVENT MODEL NOTATION

Event Models are composed of three notational elements:

Event Diagram
Concurrency Chart
Lifecycle Dictionary entries to support the above

7.2.1 Event Diagram

The Event Diagram is composed of three graphic symbols:

A rectangle containing the name of a state or its symbol

A directed line segment indicating the directional "sense" of the transition from one state to another

A label along the line segment indicating what event (if any) must occur for the transition to take place. Some transitions may take place simply because the state from which the transition will occur has completed a deterministic process and the "to" state is to be migrated to.

An example of an Event Diagram is presented in Figure 7.2.1-1.

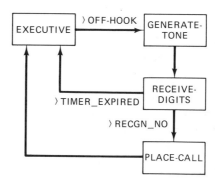

Figure 7.2.1-1:   Example of an Event
Diagram

7.3 EVENT MODEL STANDARDS

The purpose of the Event Model is to describe the real-time or state-transition-oriented characteristics of the system.

7.3.1 Conceptual Standards for Event Models:

The Event Model describes all of the states that the system may occupy.

All events or occurrences which can cause the system to change states must be described in the Event Model.

Event Models are *not* flowcharts. That is, they describe the transition(s) which can occur between states and not the details of control sequence which are described in flowcharts.

A given state described in the Event Model may include substates.

All paths through the Event Model will, eventually, cause the system to return to the EXECUTIVE or IDLE state.

All states and all events in the Event Model must be described in the Lifecycle Dictionary.

More than one state may transfer control to another state for the same reason or event.

The system may occupy more than one state.

The compatibility of states with each other is documented in a table format in the Lifecycle Dictionary.

7.3.2 Notational Standards for Event Models

No industrywide standard exists for Event Models.

The standard recommended here is to use symbology to identify states and events.

A rectangle is used to identify a state.

Each state symbol (rectangle) must be labeled with a strong (transitive) verb and optional noun.

Each state must have a minimum of one means of ingress and one means of egress.

Ingress into a state is indicated by a directed line segment which begins at the state being left and ends with the state being entered. An arrowhead appears at the end of the line segment closest to the state to be entered.

The directed line segment which describes the transition from one state to another must be labeled with the description of the event which causes the transition.

The event description must be a brief statement of an event which has occurred. For example, OFF-HOOK_DETECTED states an event detectable by a telephone system which will cause that system to change its state.

The event description label must be preceded by a > symbol.

If more than one event can cause the transfer of control to occur, the appropriate dictionary notation should be used (e.g., selection between two events).

7.4 EVENT MODEL DEVELOPMENT PROCESS

The Event Model is related to the Process Model in that the processes in that model are used as a list of tentative states in the Event Model. This loose relationship and the effects of the Event Model development are described below:

1. Utilizing the level-'0' dataflow diagram (or alternately, E-R diagram), identify the processes it contains.
2. Use the list of processes developed in step 1 as a tentative list of states.
3. Add an IDLE or EXECUTIVE state to the tentative state list.
4. Draw up a list of tentative Events by identifying all external occurrences which can cause the system to change its mode of operation.
5. Form a network of events and states drawn from the above lists. Be sure to incorporate any assumptions or observations about the operation of this network as it is developed. An example of such an observation is that a given event can occur only in conjunction with another.
6. Modify the event description(s), as appropriate, using dictionary notation wherever possible to graphically incorporate the observations made in step 5 above.
7. Walk through the network and identify any states which "split" or have substates. Split such states into two or more states.
8. Reexamine the Event Model and others and refine it as necessary.

The above process is demonstrated in section 7.5.

7.5 AN EXAMPLE OF THE USE OF EVENT MODELS [S1]

To demonstrate the derivation of the Event Model, we have selected a "smart telephone" system. The system's features are described below:

> We wish to describe the operation of an advanced type of telephone answering machine. It will have many of the features of current answering machines and a few new "wrinkles" as well. When someone calls in and the machine has been placed "on duty," it answers with a recorded message, records the message (if any) left by the caller, and hangs up. Since we do not want pranksters or unwitting callers to use up all of the message tape in a single call, the machine gives each caller only so many minutes of message time.
>
> This machine has a feature which allows the user to create specialized messages which are accessible *only* by specific individuals. When someone calls in who is checking on whether or not one of these special messages has been placed on the system for them, they type in a four-digit code on their touchtone handset when the system asks them to leave a message. When the system detects a·valid access code, it checks its message file to see if there is a message for that particular caller. If a message (or messages) is found, the system tells the caller how many messages are present and asks which, if any, the caller wishes to have played back. The caller may request replay of all messages, none, or some selected set by tapping in the user code of one or more users that messages are expected from.
>
> The caller may leave a message on the system. The message may be directed to one or more of the users who have access to the system.
>
> A subset of the users of the system (usually the owner) have other codes that they can use to change the access codes by which users can get into the system.
>
> The system can even be used to "screen" calls in the following way. When someone is actually in the office but does not want to be disturbed unless it is a call from a particular individual, they can inform the system of this, and the system will answer with the recorded message unless a person who has one of the specific codes identified to the system calls. If such a person does call, the system displays his or her name to the user, who can choose to answer it within 20 seconds or ignore it. If it is ignored, the system gives the caller whatever message(s) are waiting, records whatever messages the caller wishes to leave, and hangs up. If the user answers, then the parties can converse normally. [S1]

In examining the problem statement, we find a somewhat confusing array of roles. That is, who or what is a USER of this system? a CALLER? an OWNER? It is easy enough to get the general operating characteristics, but what about the notion of external and internal roles? Specifically, if we wish to begin by drawing a dataflow diagram, we are going to have to decide what data goes into the system and what data leaves it as well as the source of each dataflow and the destination of each. Our earlier procedure for deriving Event Diagrams encouraged us to employ a dataflow diagram. In this case, that may not be so easy to do. Let us use an alternate approach.

We will employ an Entity-Relationship diagram in order to gain the requisite understanding of this problem. We can start by drawing an E-R diagram which captures the information with which we are dealing. This diagram and its Lifecycle Dictionary definitions will help us to understand the roles and definitions well enough to derive the Event Diagram. We can begin by noting that the OWNER of this system responsible for setting up the access codes, phone lists, and special caller lists. So we can establish that:

It may be argued that INITIALIZES is not a very good relationship label, but it will do for now. Remember, this is an iterative process. Next, we may note that the OWNER may also end up calling into this system from some remote location. Hence, the OWNER may also be a CALLER. Since the OWNER may also use the system at the office location, we may wish to create a new entity called USER. A brief description of some of these terms in Lifecycle Dictionary form is presented in Figure 7.5-1.

Note that we have identified, somewhat by accident, several entities involved in this analysis which were not previously mentioned—for example, the TELE-PHONE COMPANY. These need to be incorporated into the diagram. As a result of defining these, we have also described parts of the process of using the system which are a direct result of attempting to document policy about entities. As a first-cut diagram we may wish to incorporate these elements to form an E-R diagram. If

OWNER   The person who is responsible for initiating, updating, and securing the Smart Telephone.

          OWNERS may also use the system in the same way as USERS
          OWNERS may also use the system in the same way as CALLERS
          Only OWNERS may change access or security codes

USER   A person who places calls with the system in much the same way that one would use an ordinary telephone

          A USER may also be the OWNER of the system
          Depending on physical location, a USER may also be a CALLER

CALL   Any accessing of the local or long distance telephone network through direct-dial, special long distance service, or other means.

          CALLS may be direct or forwarded
          Forwarded calls may be accomplished only by means of appropriate commands from USERS on the restricted list
          CALLS may be completed only by the TELEPHONE COMPANY

Figure 7.5-1:   Dictionary Description of Some Answering Machine Information

we note that there is a screening process at work within this system, we need to add in an entity called ACCESS CODES. This yields Figure 7.5-2.

Our earlier discussions (Chapter 4) dealt with much simpler systems. We will use this system as an opportunity to apply E-R concepts to a more difficult problem.

Figure 7.5-2 is just a first-cut at what would be a useful E-R diagram. A close examination of Figure 7.5-2 reveals that we have incorporated several undesirable characteristics into our E-R diagram. These include the fact that we have used several verbs or action terms as relationships. This makes the diagram appear to be some sort of action-oriented chart rather than a static, relational one. The figure also shows that we are considering CALLER, OWNER, and USER as separate entities when they actually share a hierarchical relationship is a result of the entity CALLER playing different roles. That is, the entity CALLER can play the role of USER. Similarly, the entity USER can play the role of OWNER of the system. Remember,

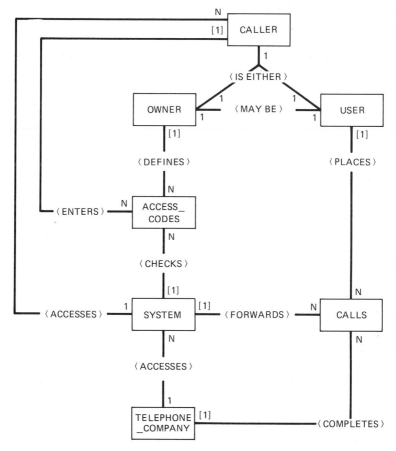

Figure 7.5-2:   First-Cut E-R Diagram [S1]

a USER has privileges which most CALLERs do not have. The OWNER may present himself or herself to the system as a USER or a CALLER.

In Figure 7.5-2, we chose to include the SYSTEM as a separate entity when it actually includes other elements on our chart. This is not a good decision in that the resulting diagram does not focus on what the enterprise is. Obviously, several modifications are in order. The revised E-R diagram is presented in Figure 7.5-3. This revised diagram allows us to focus clearly on the enterprise even to the extent that we can trace through the system and find all CALLS associated with a CALLER. It also accurately depicts the multifacetted role of the OWNER and a USER. Each may also be a CALLER. The multi-level entity in Figure 7.5-3 indicates that a USER is a special instance of CALLER and that OWNER is a special instance of USER. Note where the relationship lines contact the nested boxes for CALLER. These apply only to the relationships and entity symbols touched. For example, OWNER MAINTAINS ACCESS_CODES is unique to OWNER. A USER may not MAINTAIN ACCESS_CODES.

An alternate view of this system is the Process Model. To create it, we need to examine the dataflows which travel into and out of the answering machine. We can develop some of these flows by examining both the problem statement and the E-R diagram. We will examine the E-R diagram first. Note that the entities actually represent one of two kinds of information: (1) information which is in motion and, as a result, will be acted upon by some process, (2) information that actually represents a terminator (i.e., a source or sink of information). As such, the entities

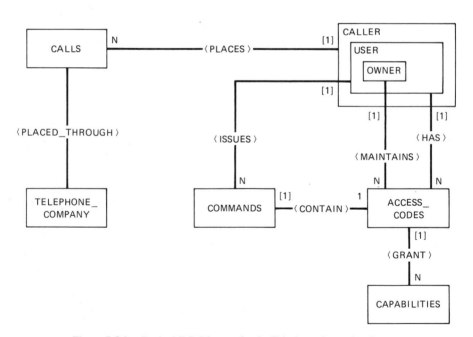

Figure 7.5-3:   Revised E-R Diagram for the Telephone Answering System

ENTITY	REMARKS
ACCESS_CODES	These are the new, revised, deleted, and required codes used to gain access to the system or make use of its features. They enter from outside the system itself and will constitute <u>dataflows.</u>
CALLER	This is a person who is either calling in from the outside or an innocent bystander. It is unlikely that this is a legitimate dataflow. We will treat it as a terminator.
CALLS	This word is used here as a noun. The functions that this system performs on CALLS is one of the attractive features it possesses. We will treat CALLS as a <u>dataflow.</u>
CAPABILITIES	Special features provided to the USER, as appropriate.
COMMANDS	User directories to the system.
OWNER	The OWNER is a person who has special privileges with respect to what features and options are used and how. It is unreasonable to consider the OWNER to be an integral part of the system. Therefore, we choose to treat the OWNER as a <u>terminator.</u>
SYSTEM	Obviously, this refers the answering machine system itself. Hence, we will treat it as neither a terminator nor a dataflow. It is, in a sense, a dummy which stands in place of the process model we will be building.
TELEPHONE_COMPANY	This is not information which flows into the system. It is a mechanism which (sometimes) sees to it that calls are actually placed. We will treat it as a <u>terminator.</u>
USER	The USER plays a role very similar to that of the OWNER in that no transformation takes place and it lies outside of the system. We will treat it as a <u>terminator.</u>

Figure 7.5-4:   Analysis of E-R Diagram to Support Process Model Development

provide information but not necessarily information which we will find listed on the E-R diagram. Figure 7.5-4 presents an analysis of our model.

Additional information needed to develop the process model may be gleaned from the problem statement, but most of what we need is already at hand. The sources and destination information have been identified. These are the terminators

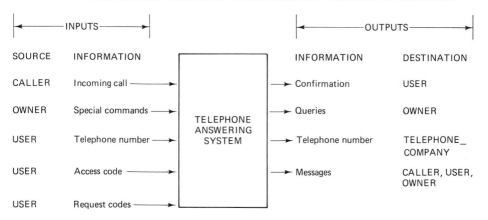

Figure 7.5-5:   Inputs and Outputs at the Context Level [S1]

TELEPHONE ANSWERING MACHINE LEVEL_0

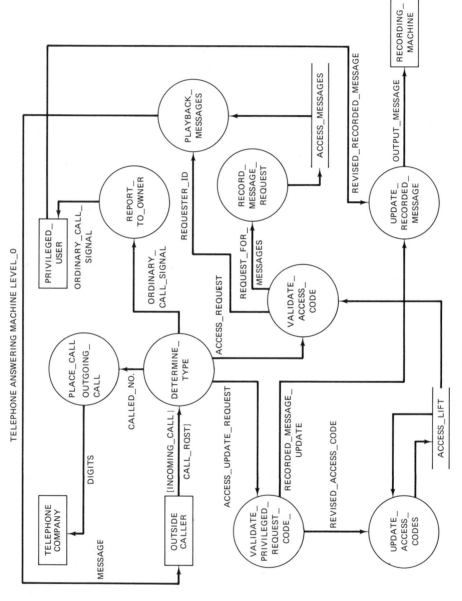

Figure 7.5-6: DFD of the Telephone Answering System [S1]

126

TABLE 7.5-1:   States and Corresponding Enter/Exit Events [S1]

State	Event to Enter	Event to Exit
Idle	On_Hook_Detected Retries_Exhausted Wait_Timer_Expired	Off_Hook_Detected Incoming_Call_Detected
Answer	Incoming_Call_Detected	Wait_Timer_Expired Incoming_Message
Record	Incoming_Message	On_Hook_Detected Messager_Timer_Expired Security_Code_Received
Command	Command_Request Off_Hook_Detected	Parameter_Update_Code Call_Digits_Received Review_Command_Received Timer_Expired Retries_Exhausted Security_Code_Received
Playback	Review_Command_Received	Command_Request Quit_Received Messages_Played + Timer_ Expired On_Hook_Detected
Call	Call_Digits_Received	On_Hook_Detected
Parameter Review/Revise	Parameter_Update_Code	On_Hook_Detected

discussed in Figure 7.5-4. Hence, our context diagram for the answering machine system would look like Figure 7.5-5. Referring back to our discussion on deriving dataflow diagrams, we can trace each of the outputs back toward the input side to form a network. This network would constitute the level-'0' diagram we need to

TABLE 7.5-2:   Reachability of States [S1]

State	Exit to	Entered From
Idle	Command Answer	Command Call Parameter Review/Revise Answer Record
Answer	Idle Record	Idle
Record	Idle Command	Answer Command
Command	Parameter Review/Revise Call Record Idle Playback	Idle Playback
Playback	Command Idle	Command
Call	Idle	Command
Parameter Review/Revise	Idle	Command

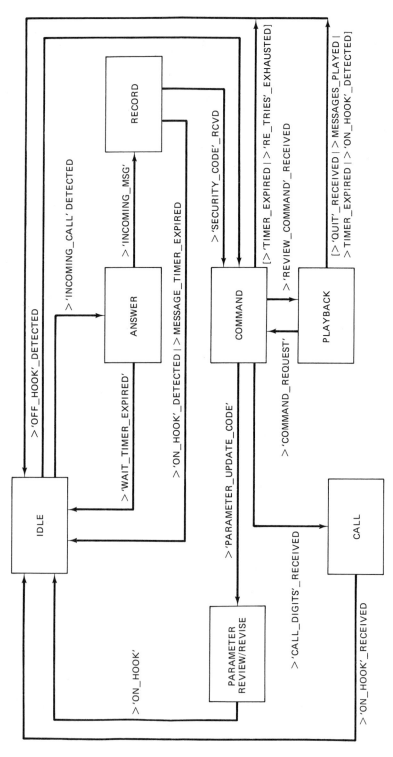

Figure 7.5-7: Event Diagram of the Telephone Answering System [S1]

derive the Event Model. Doing this will yield the DFD shown in Figure 7.5-6. Note how we have combined the requests from CALLER, USER, and OWNER into a "pipeline" for redistribution.

If we get rid of the VALIDATE and DISTRIBUTE BY TYPE processes, we have only a few candidate states left: RECORD, PLACE_CALL, and PLAY-BACK. Other states needed include IDLE, a state to field commands (which we will call the COMMAND state), and a state to enable review and revision of codes and commands (which we shall call PARAMETER_REVIEW/REVISE state). In Table 7.5-1 we list these states together with the events which will cause the system to enter and exit them. An adjunct to this listing is a reachability matrix (Table 7.5-2), which is derived from an analysis of the enter/exit events in Table 7.5-1. Combining these, we obtain a network which is our Event Model (Figure 7.5-7). Pseudocode in an alternate form is presented below for one process to demonstrate its role.

```
PROCESS: UPDATE_ACCESS_CODES
GET %REVISED_ACCESS_CODE FROM (VALIDATE_PRIVILEGED_REQUEST_CODE)
  SELECT USING %ACTION_TYPE
    WHEN %ACTION_TYPE = 'DELETE'
      PERFORM (DELETE_ACCESS_MEMBER) USING %MEMBER_ID
    WHEN %ACTION_TYPE = 'MODIFY'
      PERFORM (MODIFY_ACCESS_MEMBER) USING %MEMBER_ID AND %NEW_ID
    WHEN %ACTION_TYPE = 'ADD'
      PERFORM (CREATE_NEW_ACCESS_MEMBER) USING %NEW_ID
      AND %MBR_DESCRIPTION
  ENDSELECT
```

7.6 ROLE OF EVENT MODELS IN THE SOFTWARE DEVELOPMENT PROCESS

At first glance, Event Models may appear to have some significance for the real-time parts of software development but little elsewhere in the process. Experience with the use of Event Models has demonstrated quite the opposite. Their role is crucial for both the validation and the refinement of the partitioning of the Structured Analysis Process Model. That is, the Event Model enables us to make "discoveries" which imply changes in the Process Model's structure. This interplay between Process Model and Event Model benefits both.

REFERENCES

1. D. T. Ross and J. W. Brackett, "An Approach to Structured Analysis," Computer Decisions, Vol. 8, No. 9, September 1976, pp. 40–44.

S1. Extracted from the seminar, "Structured Analysis for Real-Time Systems," by Software Consultants International, Ltd.; Kent, Washington; Copyright 1988. Reprinted by permission.

CHAPTER 8

Structured Analysis
as a Process

Thus far, we have discussed the various tools and techniques that are a part of Structured Analysis. Now we shall describe how Structured Analysis can be planned and employed in an actual project. For a variety of reasons, the Structured Analysis phase of many projects can become a "black hole [1]." That is, it resembles the phenomenon hypothesized and to some extent "observed" in astrophysics, wherein a region in space is such that particles, even rays of light, enter it but nothing leaves.

Projects which employ Structured Analysis can experience this black-hole phenomenon when they gather and document information, refine these results, update that documentation, and so forth—but never seem to finish! Experiences with this phenomenon have shown it to stem from the following causes:

Lack of knowledge that this phenomenon can occur. Most project managers and all authors (as far as this one has been able to establish) are unaware of the dangers of giving people methods and techniques with the power and breadth of application possessed by Structured Analysis. For the first time, these people are able to collect information about the proposed effort, document it in such a way that its idiosyncrasies are obvious, its errors apparent, and its partitioning reflective of a certain quality or lack thereof. The problems come about because management, client, and the software engineers are breaking new ground. They are not used to planning such activities and, as a result, end up attempting to "Slam Dunk" the analysis.

Lack of a product-oriented plan for the analysis phase. Most management plans for the analysis phase of software development are oriented toward the classic weekly or monthly status report. This often results in the "99% complete" syndrome. That is, the analysis is reported and classified as being 99% complete for an interminable time. Often this leads management to decide that the analysis phase should be abandoned and, because the project is so far behind schedule, the design phase should be rushed through and the code "Slam Dunked" in order to bring the project in close to schedule.

The way to defeat both of the above situations is to develop a product-oriented Work Breakdown Structure which is tailored to both the project and the Structured Analysis method [1].

The primary thrust of this chapter is the development and refinement of such product-oriented Work Breakdown Structures and the presentation of experiences where groups have attempted to do this both on their own and with "expert" assistance.

8.1 THE PHASES AND MODELS OF STRUCTURED DEVELOPMENT

The model development techniques that we discuss in this book are used during specific phases. The three types of modeling techniques we have discussed relate to the development of the Process Model, the Information Model, and the Event Model. The content of each of these models is presented, by phase, in Figure 8.1-1.

PHASE	MODEL		
	PROCESS	EVENT	INFORMATION
Context Specification	— Function Model		— Data View Model
Analysis	— Dataflow diagrams — Structured English — Generic test plan — Implementation plan	— State transition diagrams	— Dictionary — ER diagrams
Design	— Structure charts — Pseudocode — Revised test plan — Revised implementation plan	— Revised state transition diagrams — Prototype	— Revised Lifecycle Dictionary — Subject Database Model
Implementation	— Structured code — Final test plan/Results — Final implementation Plan	— Process simulation — Refined Prototype	— Physical database — Implementation — Final Lifecycle Dictionary

Figure 8.1-1: Content of Models by Phase

8.2 OVERVIEW OF THE PHASES OF STRUCTURED ANALYSIS

In the classic model [2] of Structured Analysis, analysis activity is divided into four distinct phases:

> Current Physical Model Development
> Current Logical Model Development
> New Logical Model Development
> New Physical Model Development

It is recommended that these phases occur chronologically in the order in which they appear above. They can be simplified into two basic types of models:

> **Current System Model.** The goal here is to obtain an accurate description of what is currently done within the system—its inputs, outputs, processes. The goal is the some whether some sort of automated system is being replaced or we are going from a manual (or partially automated) system to an automated one.
>
> **New System Model.** The goal here is to describe the inputs, outputs, and processes that will be supported by a system which does not yet exist.

Clearly missing in this classic view is a phase that must precede model development—one in which we attempt to establish the boundaries within which the analysis and design activity will take place. This we shall refer to as the Context Specification phase. We have already added it to Figure 8.1-1.

8.3 WORK BREAKDOWN STRUCTURE FOR CONTEXT SPECIFICATION AND ANALYSIS

A Work Breakdown Structure (WBS) is a convenient way of taking a complex task, such as Structured Analysis, and decomposing it into smaller, easier-to-plan sub-tasks. In order to aid the reader in planning Context Specification and Analysis activities, we are providing a simple WBS and definitions. It should be reviewed and revised to conform to the types of systems and external requirements that you may have to deal with. Also, more detail may be needed in some environments. It is provided here for these two phases as a convenient starting point for further embellishment. The subproducts which comprise the products of each phase are also enumerated. The notation that will be used is presented in explanatory form in Figure 8.3-1. A table accompanies each of the figures that relates the subtasks, subproducts, and products. The table *briefly* describes the tasks and subtasks. For

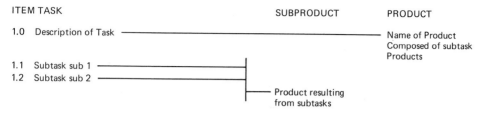

Figure 8.3-1: Notation Used to Describe WBS

purposes of simplicity, we have decomposed the analysis phase into two major subphases: Current System Model Phase and New System Model Phase.

8.3.1 Context Specification Phase [3]

This phase provides the focus for all succeeding phases. It has two key products. The System Objectives determine the overall goals of the entire effort, imply or state how success will be measured, put bounds on resources available to the project, and establish any other constraints. The Function Model specifies the major functions and processes of the real world or "external enterprise," which are related to the data included within the scope or context of the system. The enterprise (real-world) functions and processes are used during Information Modeling to determine the major "subject" databases that are to be supported by the objective system. The WBS for this phase is presented in Figure 8.3.1-1. Each element of the WBS is described in Table 8.3.1-1.

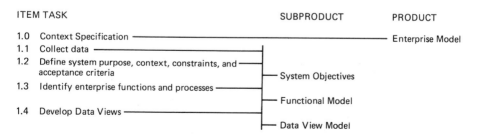

Figure 8.3.1-1: WBS for a Context Specification Phase [S1]

TABLE 8.3.1-1:   Description of WBS Elements for Context Specification Phase [S1]

1.1   Collect data

Initial data collected includes reports from opportunity identification and earlier projects related to the current effort.

1.2   Define system purpose, context, constraints, and acceptance criteria

Senior management sanctions the primary purpose of the system and defines its organizational context and constraints (budgetary, etc.) agreement with customer obtained on acceptance criteria.

1.3   Identify Enterprise Functions and Processes

An application-independent specification of the major functions and processes related to the system to be developed is created and verified with senior staff.

1.4   Develop Data Views

The enterprise functions and processes specified above determine the primary areas or views of data included in the objective system. These data views determine the areas that are refined into detailed user views during database analysis.

8.3.2 Structured Analysis Phase [S2]

The analysis phase may actually be composed of four related subphases:

Current Physical Model Development. The description of the system as it is now, including the mechanisms used to accomplish tasks (e.g., people, devices).

Current Logical Model Development. The system description in terms of functions, processes, and data with the mechanisms removed.

New Logical Model Development. The Current Logical Model with new features added.

New Physical Model Development. The Current Logical Model with the various processes allocated to automation, manual procedures, other mechanisms.

We have chosen to simplify the above by mapping them into Current and New System Models. In addition, since the physicalization of the new system is really a part of design, we have not incorporated it into this WBS. The WBS for Current System Model development is presented in Figure 8.3.2-1. New System Model development is described in Figure 8.3.2-2. The task elements of each are described in Table 8.3.2-1. Task elements are described in order of reference number.

It should be noted that the allocation process (3.10 in Figure 8.3.2-2) is one of identifying the means of accomplishment of the Process Model elements.

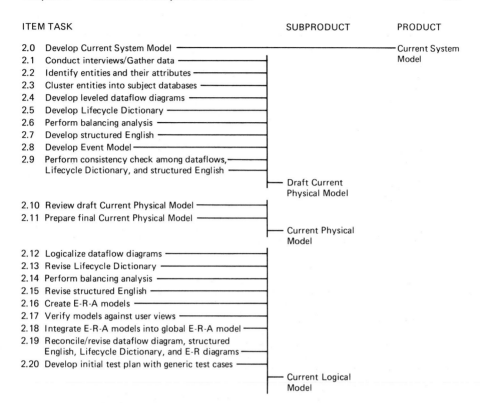

ITEM TASK   SUBPRODUCT   PRODUCT

2.0   Develop Current System Model ──────────────────────────────── Current System
2.1   Conduct interviews/Gather data ─────────────┐   Model
2.2   Identify entities and their attributes ──────────┤
2.3   Cluster entities into subject databases ─────────┤
2.4   Develop leveled dataflow diagrams ──────────────┤
2.5   Develop Lifecycle Dictionary ───────────────────┤
2.6   Perform balancing analysis ─────────────────────┤
2.7   Develop structured English ─────────────────────┤
2.8   Develop Event Model ────────────────────────────┤
2.9   Perform consistency check among dataflows,──────┤
      Lifecycle Dictionary, and structured English ──────┘
                                                        ├─ Draft Current
                                                        │ Physical Model

2.10 Review draft Current Physical Model ────────────┐
2.11 Prepare final Current Physical Model ───────────┤
                                                        ├─ Current Physical
                                                        │ Model

2.12 Logicalize dataflow diagrams ───────────────────┐
2.13 Revise Lifecycle Dictionary ────────────────────┤
2.14 Perform balancing analysis ─────────────────────┤
2.15 Revise structured English ──────────────────────┤
2.16 Create E-R-A models ────────────────────────────┤
2.17 Verify models against user views ───────────────┤
2.18 Integrate E-R-A models into global E-R-A model ──┤
2.19 Reconcile/revise dataflow diagram, structured ──┤
      English, Lifecycle Dictionary, and E-R diagrams ──┤
2.20 Develop initial test plan with generic test cases ──┘
                                                        └─ Current Logical
                                                           Model

Figure 8.3.2-1: Analysis Phase Products and Tasks in WBS Format for the Current System Model [S1]

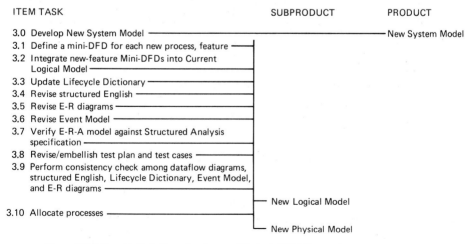

ITEM TASK   SUBPRODUCT   PRODUCT

3.0 Develop New System Model ──────────────────────────────────── New System Model
3.1 Define a mini-DFD for each new process, feature ──┐
3.2 Integrate new-feature Mini-DFDs into Current   │
     Logical Model ──────────────────────────────────┤
3.3 Update Lifecycle Dictionary ─────────────────────┤
3.4 Revise structured English ───────────────────────┤
3.5 Revise E-R diagrams ─────────────────────────────┤
3.6 Revise Event Model ──────────────────────────────┤
3.7 Verify E-R-A model against Structured Analysis   │
     specification ──────────────────────────────────┤
3.8 Revise/embellish test plan and test cases ───────┤
3.9 Perform consistency check among dataflow diagrams,│
     structured English, Lifecycle Dictionary, Event Model,│
     and E-R diagrams ───────────────────────────────┤
                                                        ├─ New Logical Model
3.10 Allocate processes ──────────────────────────────┘
                                                        └─ New Physical Model

Figure 8.3.2-2: Analysis Phase Products and Tasks in WBS Format for the New System Model [S1]

TABLE 8.3.2-1: Description of Task Elements for the Analysis Phase [S1]

2.0 Develop Current System Model

The purpose of this model is to create an accurate description of what currently exists within the scope of our study. This includes the processes, policies, and data descriptions necessary to complete this model.

2.1 Conduct interviews/Gather data

This includes other forms of data gathering about the existing system, including examination of current documentation of that system, procedures, policies, user's manuals, forms, and system architecture. The external (user) view of data that enters or is output by the objective system is obtained through interviews of persons of interest. Information sources that are examined for data use include:

Report formats
File structures
Prior data plans/analysis

2.2 Identify entities and their attributes

Through analysis of the data elements found in the user data views, generic data elements are identified. The remaining data elements are identified as attributes of the entities they characterize.

2.3 Cluster entities into subject databases

Analysis of relationship clustering and data access patterns defines groups of entities which, considered as a group, provide maximum internal contiguity and minimal external connectivity. These entity groups are the objective subject databases.

2.4 Develop leveled dataflow diagrams

The processes which describe the system are often too complex to describe completely using a single level of abstraction. In such cases, these are described in more detail at a lower, more detailed level. This enables us to control the complexity and volume of information which needs to be comprehended in order to be effective in analysis.

2.5 Develop Lifecycle Dictionary

The user data views and the E-R-A model are entered into the Lifecycle Dictionary for future use and to provide consistency and completeness analysis. This dictionary should be used as a project repository, in that all system-related information may be contained in it.

2.6 Perform balancing analysis

A potential drawback of using leveling to control complexity is that "discoveries" and errors may occur at the lower levels of detail during analysis. This means that the lower, detailed versions of the system may become inconsistent or incompatible with the upper, less detailed ones. Balancing is the process of ensuring that the data that entered the upper-level process is conserved when that process is expanded into a more detailed version. The only way to establish this is through manual inspection combined with the use of an accurate, up-to-date Lifecycle Dictionary.

2.7 Develop structured English

Structured English (SE) is an Englishlike statement of how each process accomplishes its task. Pseudocode is often used to accomplish this. Two approaches are employed in the development of SE. One is to write SE for every process. The other is to write it only for the lowest-level processes. Either approach may be used; the first is recommended. The SE would also be included in the Lifecycle Dictionary.

2.8 Develop Event Model

Using the processes contained in the level-'0' dataflow diagram as a set of candidate states, form a state transition diagram. The events or occurrences which can cause the system to transfer from one state to another are found through an inspection of the interviews, policy statements, and general knowledge of the operation of the system being studied. Each state is a potential candidate for leveling if the process from which it was derived is also leveled.

2.9 Perform Consistency Check among dataflows, Lifecycle Dictionary, and structured English

The objective of this activity is to ensure that all data which appears on the dataflow diagrams is defined in the Lifecycle Dictionary, and all items (e.g., data, processes) defined in the dictionary are actually used in the model. That is, all information associated with the model is defined consistent with all parts of the model, and all information that is defined is used in the model.

TABLE 8.3.2-1: (continued)

2.10   Review draft Current Physical Model

This is for purposes of revision, completeness checks, and correctness.

2.11   Prepare final Current Physical Model

Integrate all elements that comprise the Current Physical Model into a deliverable document.

2.12   Logicalize dataflow diagrams

This process involves the removal of all references to people, places, things, form numbers, colors, and other implementation-oriented characteristics that may be present in the Current Physical Model. The intent is to retain the *function* that all of these things perform without retaining the *means* by which they were accomplished.

2.13   Revise Lifecycle Dictionary

The subject database specification (including the E-R-A model) is entered into the Lifecycle Dictionary and put into a "frozen" status for use in guiding database design, implementation, and maintenance efforts.

2.14   Perform balancing analysis

See item 2.6.

2.15   Revise structured English

The removal of the mechanisms by which tasks are accomplished will require the review of all structured English (SE) to ensure that little or no reference is made to implementation-oriented information. This will necessitate the creation of several new entries in the Lifecycle Dictionary.

2.16   Create E-R-A models

Use of Entity-Relationship-Attribute (E-R-A) diagramming techniques provides a diagrammatic specification of the relationships between entities, and of the attributes that specify characteristic information about the entities. Each E-R-A diagram may correspond to one or more data view specifications.

2.17   Verify models against user views

Because E-R-A modeling combines information related to one or more user views, the database modeler must verify that each data view can be supported by the refined E-R-A model.

2.18   Integrate E-R-A models into global E-R-A model

For each subject database, the E-R-A diagrams involving entities in that subject are integrated into a global E-R-A model for the subject database. In complex data modeling situations, there may be a need to partition the global E-R-A model through use of several submodels for a subject database. An entity appears in one and only E-R-A model.

2.19   Reconcile/revise DFD, SE, LD, and E-R diagrams

Again, this is a form of confirmation to ensure that all elements associated with the model are accurate and consistent with each other.

2.20   Develop initial test plan with generic test cases

This involves the creation of a process for establishing that the system, when implemented, actually performs the tasks for which it was initiated. Generic test cases (i.e., test cases describing what will be tested for without identifying the data values) are outlined at this point; later in the development process they can be more accurately described.

3.0   Develop New System Model

This involves the adding of new features to the logical model of the existing system and the updating of all supporting documentation.

3.1   Define a mini-DFD for each new process, feature

This is similar to item 2.2 but relates to the incorporation of new features into the system model.

3.2   Integrate new-feature mini-DFDs into Current Logical Model

This incorporation is for purposes of completeness.

3.3   Update Lifecycle Dictionary

See item 2.4.

(continued)

TABLE 8.3.2-1: (continued)

3.4   Revise structured English

See item 2.6.

3.5   Revise E-R diagrams

See item 2.16.

3.6   Revise Event Model

See item 2.8.

3.7   Verify E-R-A model against Structured Analysis specification

At this point the E-R-A model is integrated with the Structured Analysis process model to verify that the model supports that "user view" also. The Structured Analysis model is then modified to reference only subject databases and their entities, attributes, and relations. Any databases created during Structured Analysis are now incorporated into the Information Model.

3.8   Revise/embellish test plan and test cases

This is a natural outgrowth and refinement of the activity supporting item 2.20.

3.9   Perform consistency check among dataflow diagrams, structured English, Lifecycle Dictionary, Event Model, and E-R diagrams

See item 2.19.

3.10 Allocate processes

This involves identifying where each process will be performed, new/existing hardware or software.

REFERENCES

1. Peters, L. J., "How to Prevent Structured Analysis from Becoming a 'Black Hole,'" Structured Development Forum VIII, Seattle, Washington, August 1986. Copyright Software Consultants International Ltd., 1986.

2. T. De Marco, *Structured Analysis and System Specification*. New York: Yourdon Press, 1978.

S1. Extracted from the seminar, "Structured Analysis of Real-Time Systems," by Software Consultants International, Ltd.; Kent, Washington; Copyright 1988. Reprinted with permission.

S2. Extracted from the seminar, "Managing Structured Projects," by Software Consultants International, Ltd.; Kent, Washington, Copyright 1987. Reprinted with permission.

PART III

Structured Design

Structured Design is directed at providing the software engineer with a means of developing cost-effective, maintainable systems. Nearly everything about it is directed at this goal. Its primary notation, the Structure Chart, is architectural in its view. Its design quality evaluation criteria encourage simplicity and lucidity of concept. The heuristics incorporated into it discourage shortcuts and imprudent programming practices and encourage practices "good" programmers recommend [1].

Structured Design is the result of a desire on the part of its authors to formalize what they consider to be good programming practice. In their view, ease of maintenance is synonymous with quality systems. A beneficial side effect for the software engineer is that what is easy to maintain is also relatively easy and inexpensive to construct. This is not surprising, since nearly every high-technology industry has already made this discovery. Thus, Structured Design is a software design method which closely parallels the current trend in technology toward systems which are at once easy to maintain and inexpensive, or at least cost-effective, to construct.

The lofty goals of Structured Design are accomplished through the consistent implementation of specific philosophical views. One of these, functionality, is stressed throughout the practice of Structured Design. At the heart of this method is the notion that a program, a group of programs, or a group of systems is nothing more than a collection of functions. Once again, this harks back to the concept of the "black box" used in Structured Analysis, electrical engineering, and just about everywhere else. In essence, the software designer must at first "see through" the programs, modules, and routines and examine relationships. One must temporarily forget about how the system at hand might be implemented and treat it as a collection of abstractions—logical functions. This gives the software designer maximum freedom in examining alternative system architectures—a crucial element in successful software design practice [2]. The mapping of logical functions into physical modules is delayed until the late stages of the design effort.

Perhaps the outstanding unique aspect of Structured Design is that it includes several different ways of evaluating a design. Note that the concepts we refer to here are applicable to the practice of design itself, not just software design. Another unique feature of Structured Design is that the design quality criteria incorporated into it may be applied to *any* software design, not just those produced using this method. Although literally dozens of software design methods are available today [2], only one, Structured Design, incorporates a design evaluation scheme which can be applied to existing software as well as designs that were developed using some other software design method.

This situation has several interpretations. The one most likely is that certain kinds of rules or principles of software have widespread, almost universal application. They practically constitute some sort of "natural law" [2] of software design. Once software engineers become familiar with Structured Design's evaluation criteria, they exclaim, "That's obvious, simple, easy!" or "Sure, I have always practiced that." But have they? With Structured Design, much as with religion, the enunciation of the word is easier than its practice. Thus, Structured Design's evaluation criteria challenge software engineers in three ways:

The suppression of coding issues, making these issues subservient to design principles.

Acceptance of the systems view and not the local view as the key indicator of success or failure.

Resolution of issues related to "finishing" the design through application of a predetermined value system.

Structured Design provides two effective means of evaluating software design decisions. One measures the relative effect of the decision on module quality, while the other measures the relative effect of the decision on the relationship(s) between modules. Note that we are dealing with *relative* degrees of module simplicity or module relationship complexity, *not* some absolute measure. The main value of such a systematic evaluation philosophy lies in providing the software design engineer with an objective and impersonal means of evaluating the alternative design characteristics that are available. A side effect is the consistency brought to the software systems that are produced. It greatly reduces the variability between structure and content. This, in turn, will greatly reduce maintenance costs, since those maintaining the software will repeatedly see the same or nearly the same patterns from one system to the next. A primary cause of maintenance problems is the individuality of programming styles and system organizing. Those are a result of variations in the value systems of individual software engineers. The use of a consistent evaluation scheme greatly reduces this.

III.1 RELATIONSHIP BETWEEN STRUCTURED ANALYSIS AND STRUCTURED DESIGN

A big advantage of using Structured Analysis as a precursor to Structured Design is that nothing developed during the analysis goes to waste. Each of the major deliverables from Structured Analysis (i.e., dataflow diagrams, data dictionary, structured English, and entity-relationship diagrams) is used in the Structured Design phase to develop one of its major deliverables (i.e., structure charts, design data dictionary, pseudocode, and database design) as depicted in Figure III.1-1. Each of the deliverables in Structured Design will be discussed in detail in the chapters which follow.

Just as "A statement of the problem is a statement of the solution" [3], the results of requirements definition basically shape the design. This is nowhere more evident than in the relationship between Structured Analysis and Structured Design. A poorly partitioned, unrefined Structured Analysis is going to leave the software designer who uses Structured Design with a lot more work to do than one might anticipate. Worse, this situation can result in a design and system whose structure or "shape" does not parallel or closely approximate the relationships present in the problem. The result may be a delivered system that requires considerable change to meet the client's needs. A poor analysis job is going to cause problems, whether it was done by the same person doing the design or not. Some of the most common failings during Structured Analysis are shown in Figure III.1-2 together with the symptoms they produce during design.

The completeness and accuracy of the problem model will determine the amount of work to be done in design to obtain a system blueprint which is at least feasible to implement. A unique characteristic of Structured Design is that it includes several types of software design quality criteria, by which the software engineer can objectively determine the relative quality of the design on both the

STRUCTURED ANALYSIS PRODUCT	used to produce	STRUCTURED DESIGN TOOL
Dataflow Diagrams	→	First-cut structure chart
Data Portion of Lifecycle Dictionary + flags	→	Design data Dictionary
Pseudocode + implementation aspects	→	Pseudocode
E-R diagrams + database characteristics	→	Database Design
Event diagrams	→	Leveled/refined Event Model

Figure III.1-1: The Derivation of Structured Design Deliverables from Structured Analysis Deliverables

Problem	Effect on Design
— Dataflow diagram not balanced	Interface problems
— 'Fuzzy' process names	Need for repartitioning
— No dictionary	Duplicates and aliasing
— Dictionary not correlated with dataflow diagrams	Duplicates, aliasing, and interface problems
— Dataflow diagrams not correlated with dictionary	Duplicates, aliasing, and interface problems
— Pseudocode not validated	Logic errors
— Pseudocode not correlated with dictionary	Data aliases, interface problems

Figure III.1-2:   Problems Poor Analysis Can Cause in Design

microscopic and macroscopic levels. The final or "to be implemented" design is the result of repeated application of the quality criteria to the design, starting with the initial design resulting from a translation of the analysis (Figure III.1-3). The Structure Chart is the primary focus of attention in many discussions about Structured Design. Without the pseudocode to describe what each module is to do and the data dictionary to define what the data transmitted between and used by modules is, Structure Charts are not very useful.

For many years software engineers have complained that a stronger tie is needed between requirements and design [4]. As we have seen, Structured Analysis and Structured Design have effected that close relationship. As a reminder, our reference chart is presented in Figure III.1-4.

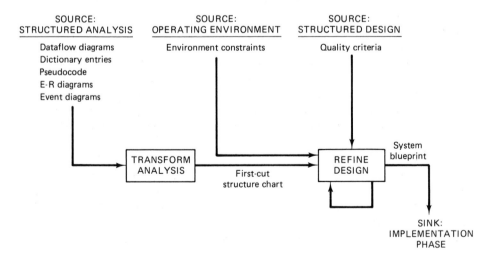

Figure III.1-3:   Derivation of the System Design (Blueprint) Using Structured Design and Structured Analysis

	DEVELOPMENT PHASE		
TYPE OF MODEL	STRUCTURED ANALYSIS	STRUCTURED DESIGN	IMPLEMENTATION
PROCESS	Dataflow Diagrams Dictionary[1] Pseudocode	Structure Chart Dictionary[2] Pseudocode[4]	Code Dictionary[3]
INFORMATION	Entity-Relationship Diagram Dictionary[1]	Database Design Dictionary[2]	Implemented Database Dictionary[3]
EVENT	Event Model[5] Dictionary[1]	Event Model[6] Dictionary[2]	Queuing Model Dictionary[3]

[1] This dictionary includes data definitions: pseudocode for each process in the analysis model; descriptions, attributes, and content for the entities and relationships in the E-R diagrams; and descriptions of the events and states in the Event Model.

[2] This dictionary contains everything found in the analysis version of the dictionary plus definitions of all flags and pseudocode for all modules that were added as part of the design process or not otherwise present in the analysis.

[3] This dictionary has been updated to incorporate all information developed during the design activity which has been modified as a result of implementation considerations.

[4] The pseudocode referred to here incorporates any and all flags employed in the Structure Chart. All pseudocode is contained in the dictionary as well. Much of this may be identical to or based on the pseudocode that was developed during the analysis.

[5] Event Model refers to a simplified form of state transition diagram.

[6] This Event Model would be further refined and leveled based on design "discoveries."

Figure III.1-4:   Information Flow by Phase and Model Type [S1]

III.2 VIEWPOINT

Among the views of a software system employed by the various design methods, the Structured Design view may be unique [2]. It enables the software designer to treat the system as an abstraction and supports this view with several guidelines, goals, and heuristics. One advantage this abstract view confers is the ability to consider design decisions on their own merits and not on the basis of what one individual or another considers "good" coding practice. Yes, coding practice. Coding often dominates discussions related to design as well as private contemplation about it. Since coding is something of a deterministic process and most of our education during formative years was related to deterministic problem solving, it is natural that most people find design discussions disquieting and coding discussions at least invigorating. But this can cause problems in software systems development. Such premature consideration of implementation leads to poor, unmaintainable designs [5].

   In Structured Design, implementation issues are forestalled until near the end of the design process. This permits the software engineer to propose and consider many alternative designs and to select the best from among them. Software engineers have related that this can represent a significant time savings. Recently, one engineer noted that his design team had refined the architecture of a particular system a total of five times, each version successively better than its predecessor, in a period of about an hour, but if it had been coded, such refinement would have taken several times as long or would not been done at all, owing to the complexities involved with code.

The need to consider alternatives before committing to an implementation has long been recognized as the hallmark of sound design practice [6], but the problem remains of determining whether or not one design is an improvement over some previous proposed design. In this situation, Structured Design not only helps provide a guideline for the designer to follow, but it also affects the way in which the design process proceeds. The ability to consider many alternative system architectures, however dull or radically different, is greatly enhanced and supported by Structured Design. Also, the unique value system incorporated into Structured Design provides a framework within which the software designer can not only evaluate alternative designs, but establish whether or not successive refinements are worthwhile. Since there is no stopping rule in design activity [2, 7] objective, relative quality evaluation is the next best thing. Structured Design differs from the other software design methods, in that they tend to concentrate effort on getting from a specification to an implementable design in such a way that alternative architectures are either disallowed or tacitly ignored, while Structured Design encourages and supports the consideration of as many alternative designs as the design team can muster.

Structured Design places a great deal of emphasis on issues associated with both the partitioning of the system's tasks into groups of subtasks (which may or may not become physical modules in the implemented system) and the relationships that exist between and among those tasks. Deemphasized are the procedural or sequential aspects of the system. This is in keeping with the emphasis on low construction and maintenance costs. Why? Consider this: The serious maintenance problems experienced with software systems have certain common characteristics. Think about the worst maintenance problems you personally have experienced or heard about. When they were finally resolved, were most of them caused by some sort of sequential or procedural problem (e.g., performing two computations out of order) or were they the result of some obscure interaction or interface problem (e.g., one module clobbering the part of a COMMON BLOCK used by another module)? A personal survey of several hundred software development and maintenance personnel indicates that the overwhelming majority (9 out of 10) of the really traumatic maintenance problems stemmed from interface difficulties. These problems were troublesome but solvable. How were they solved? These same people indicated that their solution approach was to trace the data through the system to identify how things were "messed up." What they used may be most concisely referred to as "dataflow analysis."

The use of Structured Analysis prior to Structured Design reduces interface problems by basing system partitioning on dataflow analysis. Hence, the emphasis in Structured Design on architectural issues, the population of modules, and their interactions is, in a way, a productivity aid in that labor resource is focussed where it will do the most good. This emphasis tends to collide head-on with the views held by many who consider themselves software designers. Many people tend to get into arguments regarding "efficiency," database access methods, record layouts, and print formats early on in the design phase. In fact, these issues are secondary to architectural ones. Although the architectural issues may require more concentration, judgement, and contemplation, the payoff for the designer and the client can be quite high. For example, which is a better system, one which reflects great coding or implementation practice but is expensive (i.e., cost ineffective) or impossible to maintain and expand, or one which is cost effective to maintain and expand but reflects only average skill at coding? Certainly, there are situations in which we can have both characteristics. However, given that most systems will undergo nearly continual change, refinement, and extension over their lifespan, we, as software engineers, need to consider ease of maintenance as our primary goal. Reduced development cost turns out to be a side effect of this approach.

III.3 ORGANIZATIONAL IMPACT

Perhaps the most overlooked feature of Structured Design is its effect on the people, groups, and organizations which use it. Since design itself is such a poorly defined and understood activity and software engineers come from so many different disciplines, it is not unusual for walkthroughs, reviews, and one-on-one discussions regarding one design feature or another to break down into arguments involving personal opinion and "one-upmanship." Each person approaches the problem of software design from his or her own Weltanschauung or world view [2, 3]. Since we, as software engineers, come from so many diverse backgrounds (i.e., we have no common training, degree, or discipline), our views are divergent. The introduction of Structured Design can help to reduce that divergence by providing all concerned with a common value system—that is, a uniform view of what is or is not desirable in a system.

Introducing Structured Design into an organization which has few, if any, documented and practiced policies and procedures may not bring about the kind of change that management and technical personnel had intended. In such organizations there may be a backlash of sorts, resulting in a split between those who embrace this new method and those who reject it. Often, training and more training are seen as the answer. But training alone will result in little or no positive effect other than to provide education firms with a lot of revenue [8]. So what can be done to ensure that the introduction of Structured Design (and probably Structured Analysis) has as positive a benefit as possible for the organization? The approach differs from organization to organization [8], but some common guidelines are evident:

> Make it clear from the start that Structured Design is not a panacea for all the problems being experienced in the software development organization. Most of these are related to management and not technical issues [8].
>
> Obtain support from all who may be affected by this change. The people most reluctant to make the transition to Structured Design are (perhaps surprisingly) in management—not in the technical part of the software engineering arena [9]. However, if they edict the adoption of this method without gaining something approaching a consensus, those who were not part of the decision will not support it, thereby compounding organizational problems [8, 10].
>
> Emphasize the flexibility that Structured Design provides and downplay the misconception that it is restrictive and denies the software designer freedom of expression. It is an advisor, *not* a dictator.

None of these guidelines has been adequately addressed by proponents of Structured Design. But these points can make the difference between success and failure in its use. In this text, we will concern ourselves primarily with the technical issues. The management aspect is discussed in more detail in other articles and texts [8, 9, 11].

The positive side of all this is that organizations who have successfully made the transition to Structured Design have experienced more effective reviews, walkthroughs, and improved feedback and estimating on projects. Certainly, problems remain, but the effect that Structured Design (and Structured Analysis) can have on human communications is a positive one and well worth the effort.

III.4 STRUCTURED DESIGN AS A PROCESS

Structured Design development passes through two primary phases: Logical Design development and Physical Design development. Of the two, Logical Design requires more flowtime and effort. The goals of these phases are very different. Logical Design transforms the dataflow diagrams developed during the Structured Analysis phase into a *first-cut design,* embellishing and refining this design to reflect the kinds of quality attributes that Structured Design encourages. The Physical Design phase involves the modification, as necessary, of the Logical Design to accommodate whatever changes are necessary due to the operating system the software will be running under, the hardware it will execute on, the programming language(s) it will be written in, and timing and sizing considerations.

Two separate phases are needed because the two activities are very different and require a different mind set. Logical design is largely a left-brained [12] activity in that it encourages the examination of alternatives without regard for the physical or runtime characteristics of the system. What we are after is a maintainable, well-structured system. This requires intuition and imagination, *not* the kind of sequential, syntactic, and semantic thinking we engage in when using a programming language. Physical design does not involve nearly as much intuition and imagination, since the "rules of the game" are clearly defined by the programming language and operating system as well as other factors.

Looking at design as a process, we see that it passes through a series of stages [1]. There is no widespread agreement on the number and nature of these stages. One useful model for design describes four stages [13]. If we adapt them to Structured Design, we obtain five steps. The first four constitute the Logical Design phase, the last the Physical Design phase. These steps are described below:

1. Transformation of the problem model into an initial solution model. This procedure is described in detail later in the book. When Structured Analysis is used to define the problem model (specification), the procedure is relatively straightforward. It amounts to a translation more than anything else. When the specification approach is something other than Structured Analysis, considerably more is involved; in fact, the specification must be put into the form of a Structured Analysis. In either case, the result of this step is strictly a "first-cut" Structured Design. A great deal of refinement and improvement must be employed before a design is obtained which would be prudent to implement.

2. Embellishment of initial design to include necessary features. The initial design from step 1 will not contain all error-processing modules, modules which interface to external devices, reporting modules, and others. In this step, these types of modules are added in to complete the overall system function set. At this point we have a design which is complete but would not be very useful to implement. We have not evaluated or refined it in any way.

3. Transformation of the structured English into pseudocode. If structured English was developed during the analysis phase, it must be transformed into pseudocode. Using pseudocode during analysis saves this extra step. Either way, the process descriptions that we have at this point describe what only *some* of the modules in the design do. They do not describe them all, because some modules do not have corresponding processes in the analysis. Hence, pseudocode must be written for these "new" modules. In addition, the structured English that describes a process is oriented toward a statement of policy and is *not* intended to be code oriented.

4. Evaluation and refinement of the qualities of the design. Utilizing the procedure previously described, the software designer evaluates the nature of the interactions that occur between pairs of modules (coupling), the nature of the modules themselves (cohesion), and other complexity and heuristic criteria.

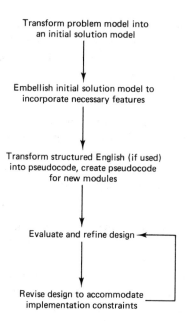

Transform problem model into
an initial solution model

Embellish initial solution model to
incorporate necessary features

Transform structured English (if used)
into pseudocode, create pseudocode
for new modules

Evaluate and refine design

Revise design to accommodate
implementation constraints

Figure III.4-1:   Overall Flow of the Development of a Structured Design

5. Revise the design to accommodate implementation constraints. We enter this step with a software design which has been evaluated, refined, embellished, and revised. What we have is a high-quality abstract or logical design. Although it is considerably better than the one we started with in step 1, it is still not necessarily a design which would be prudent to implement. In this step we make revisions to accommodate timing and sizing constraints, availability of existing or utility software, and interfaces to database management systems and to take advantage of specialized language features and other implementation realities.

These five steps represent a general flow of the process and by no means describe the details associated with the nearly continual refinement of a design from its inception through implementation. A dataflow representation of this process is presented in Figure III.4-1.

III.5 SUMMARY

We enter the Structured Design activity with a statement of what the system must do in the form of a Structured Analysis. We will leave design with an exact blueprint of what will be built *and* a basis for the configuration and content of that blueprint. Just as in the case of construction industry, it is this blueprint which will be modified and maintained over the life of the system.

The notation modifications shown in this section are supported by an automated tool [14].

REFERENCES

1. B. W. Kernighan and P. J. Plauger, *The Elements of Programming Style*. New York: McGraw-Hill Book Company, 1974.

2. L. J. Peters, *Software Design: Methods and Techniques*. New York: Yourdon Press, 1981.

3. H. W. J. Rittel and M. M. Webber, "Dilemmas in a General Theory of Planning," Institute of Urban and Regional Development, Working Paper No. 194, Berkeley, University of California, November 1972.

4. L. J. Peters, "Managing the Transition to Structured Programming," *DATAMATION*, May 1975.

5. E. Yourdon and L. L. Constantine, *Structured Design*. New York: Yourdon Press, 1975.

6. A. B. Saarinen, ed., *Saarinen on His Work*, rev. ed. New Haven, Ct.: Yale University Press, 1968.

7. L. J. Peters and L. L. Tripp, "Is Software Design Wicked?," *DATAMATION*, Vol. 22, No. 5 (May 1976), pp. 127–36.

8. L. J. Peters, "The Chinese Lunch Syndrome in Software Engineering Education: Causes and Remedies," IEEE Workshop on Software Engineering Technology Transfer, Miami, April 1983.

9. L. J. Peters, *Making the Transition to Structured Design*. Auerbach Publishing Company, 1980. N.J.

10. W. Ouichi, *Theory Z*. Reading, Mass.: Addison-Wesley, 1981.

11. D. Couger and R. A. Zawacki, "Key Factors for Motivating Computer Professionals," in *Systems Analysis and Design: A Foundation for the 1980's*, ed. W. W. Cotterman et al. New York: North-Holland Publishing Company, 1981.

12. L. J. Peters, "How to Prevent Structured Analysis from Becoming a 'Black Hole'," Structured Development Forum VIII, Seattle, Washington, August, 1986. (Copyright 1986, Software Consultants International, Ltd.).

13. J. C. Jones, *Design Methods*. New York: Wiley-Interscience, 1970.

14. *Advanced Structured Analysis and Design Package Level Dictionary*, Kent, Washington: Software Consultants International, Ltd., 1987.

CHAPTER 9

The Structure Chart

The Structure Chart is probably the most widely recognized tool incorporated into Structured Design. It may also be the most widely misused and misunderstood. The Structure Chart is intended as a means of depicting the architecture of a system in much the same way that a blueprint depicts the architecture of a home. Just as the blueprint shows the rooms and their relationships to one another, the Structure Chart shows the module population and their relationships to one another. Similar to the blueprint, it does this without indicating which are likely to be used most, which to be used first or last, or the order of their use. Hence, the Structure Chart shows the overall population of modules making up the system but not the order in which they execute, their size, execution time, and other operational features. All the Structure Chart depicts is the population of modules, their position in a hierarchy of calling and called (or invoked) modules, and the data they exchange.

This information may not sound too useful, but many maintenance programmers wish that it existed for the systems they are responsible for. During development, this information is compact enough to focus the software engineer's attention on the larger, system-level issues associated with the design. This helps to keep designers from getting caught in the maze of detailed, implementation-oriented, local issues. Owing to the psychological makeup of most software engineers [1] and the general nature of the detail-oriented work that programming is, there is a strong tendency for people to get bogged down in such issues. For example, software design reviews are held, primarily, for the purpose of evaluating, refining, and generally disseminating information about the design of modules and systems.

However, it is common for discussions to break out during design reviews which focus around the relative merits of one programming language or feature versus another, or efficiency, or timing and sizing.

Doesn't this mean that many people are missing the point of such reviews? Yes, perhaps, but there is a simple explanation for this phenomenon. Design and analysis are done first and foremost by people, and people have a tendency to feel more comfortable dealing with immediate, local issues (like what to have for breakfast) rather than larger, more global ones (such as what an appropriately balanced diet would be for a person of a certain age and activity level). In order to amplify this concept, we pose the following question: "Which would be easier and less costly to develop and maintain, a system composed of elegant modules arranged in some complex, perhaps undefined way, or one composed of simple modules organized along a rational set of simple, consistent rules?" Obviously, stated in this way, the latter would be the more desirable of the two. But do we really make this choice in practice? Probably less than we should. We still read software engineering journals which publish articles on the speed of one algorithm over another, or on how to save space using certain kinds of programming techniques, automated programming, error-free code, an improved programming language, and other microscopic aspects of a system. These are all valid considerations, but we must bear in mind that the systems that we are designing and building are probably subject to more change than any of the other systems that our client is using. Speed, elegance, and error content mean very little to the client if the system cannot be cost-effectively modified to meet the changing demands of the business environment.

9.1 CONCEPTS INCORPORATED INTO STRUCTURE CHARTS

Structure Charts enable the software engineer to graphically depict the architecture of software systems. They show the relationships which exist between and among modules and externalize with respect to design-related issues. Hierarchical (i.e., one module calling another module), communicational (i.e., a module transmits some data elements to another), and synchronization (i.e., non-hierarchical coordination systems) relationships are described. The graphics embodied by Structure Charts display the population of modules in a system and their interactions. Other characteristics or features of the design that can be depicted will be described later.

Since some processes are needed for implementation that were not included in the analysis, the Structure Chart includes many modules whose primary jobs include the documentation of errors, the provision of an interface with external devices, and the performance of tasks which were "discovered" during the transition from analysis to design—validation-oriented modules, modules needed to compute an intermediate result, and modules required to support requirements that were not part of the original specification but were put in at some late point in the development cycle.

Structure Charts employ several ideas that have become inherent in the movement of software engineers toward ''structured'' software technology. These ideas include hierarchical decomposition, modularity, functionality, standardized interfaces, and simple relationships between modules. While proponents of other methods have often emphasized one or more of these as the key to successful software design, Structured Design utilizes a synergistic combination of all of them. Nowhere is this more evident than in the Structure Chart. We will demonstrate this by starting with a simple diagram and introducing each idea both graphically and conceptually. Concepts most often incorporated into Structure Charts include:

Hierarchy. The idea that some things or beings are subservient or subordinate to others may well be one of the oldest abstract concepts employed by humans to cope with their surroundings. Fortunately, the software community rediscovered hierarchy some years ago and has been using it ever since. The Structure Chart employs hierarchy in a twofold way: (1) to depict certain modules as being in control of or being controlled by other modules, (2) to place restrictions or suggest limits on how many modules a ''boss'' (or superior) module may control. We will delve into such rules later in this chapter.

Modularity/functionality. The idea of a module apparently grew out of a need to change systems more easily. The basic notion was to collect together a set of instructions which belong together (according to some rule of thumb) and let them be addressable or invokable via some sort of name or label. Sounds a little ''fuzzy'' or inexact, doesn't it? Well, it is just that—*fuzzy*. This inexactitude has led to the proposal of numerous approaches to modularization. For example, articles in the late sixties proposed that systems be modularized in such a way that all I/O was handled by one module, all validation by another module and all error handling by a third. This was a natural outgrowth of the trend toward centralizing everything that was taking place in data processing at that time. We are still feeling the effects of that era. What was and still is causing the software designer difficulty is that given a system with many different functions or tasks to perform, what would be a prudent arrangement of those tasks such that our ability to construct and maintain them will be enhanced? Each of us has our own *Weltanschauung* (i.e. world view, philosophy of life) with which we organize and arrange the myriad of information presented to us. This diversity is definitely healthy in private life but in the software profession it leads to systems that only the author can comprehend. Structured Design, via the Structure Chart, provides us with a mechanism with which to temper, not eliminate, diversity. Through the use of rules and advisories regarding the use of the Structure Chart, many of the problems of modularizing systems become controllable. The viewpoint that predominates the use of Structure Charts is that they are populated by modules and modules are functional in their nature. That is, functions and modules are one in the same.

Protocol. A serious problem encountered by those who maintain software systems is that there is little or no consistency in how modules relate to one another. Just the idea of eventually always returning system control to the same point in the chart would help many existing systems. Structure Charts employ a simple set of rules regarding the nature of the protocol between calling and called modules. It is that whenever control is relinquished by a superior module to an inferior one, control eventually returns to the superior module. Think about the worst system you have ever had to make a significant change to. What would have been the effect of employing such a rule?

Explicitness. Structure Charts seem to support the premise that software designs should be "X"-rated. That is, they are explicit in many significant ways, and they leave nothing of any significant architectural impact to the imagination of those who must review or implement the design. It is possible to use obscure interfaces when employing this or any other software design method, but that obscurity is much harder to achieve with the Structure Chart than with other notations. When one is inspecting a Structure Chart with such a masked interface, the questions that need asking become obvious.

Information Communication. The communication of information in the Structured Analysis phase was accomplished by means of pipelines called dataflows. In Structured Design, the dataflows which linked processes now become data couples which pass between calling and called modules. Hence, if two modules need to pass information between them, one must call the other. There are some alternatives to this rule, which we shall discuss in later chapters.

Synchronization. The coordination of the activities of two or more modules without using the CALL and parameter passing technique. The need for this type of relationship has grown tremendously since Structured Design was first published. This need is the result of the increased use of real-time systems in defense and commercial work.

At this point, many doubts about the advantages of Structure Charts over other notational schemes should have been put to rest. As we proceed with this chapter, the integral role that the Structure Chart plays in Structured Design should become more obvious. It is not a coincidental notation but rather one that encourages, if not forces, its users to employ at least some minimum level of quality in their software designs.

9.2 NOTATION

The Structure Chart utilizes four primary symbols (Figure 9.2-1). These correspond to the MODULE, the CALL or invocation of one module by another, and the two

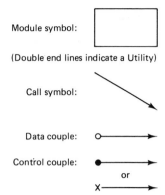

Module symbol:

(Double end lines indicate a Utility)

Call symbol:

Data couple:

Control couple:

or

Figure 9.2-1:   The Primary Symbols Used
in Structure Charts

kinds of information that can be sent or communicated by one module to another—
data and flags. Several other symbols are used to describe specific conditions and
system properties. These will be discussed in later chapters. Before we proceed to
detail the use of Structure Charts, a few definitions and examples of the use of the
primary symbols are in order.

9.2.1 The Module

In Structured Design, a *module* is defined as a set of instructions which can be
invoked by name. To illustrate this in more familiar terms, examples of modules in
several programming languages are presented in Table 9.2.1-1. Notice that the key
distinction between, say, a set of programming statements and a module is that the
contents of a module can be referred to collectively under a single label. Presum-
ably, that label tells other software engineers just what that module does—nothing
more and nothing less. Although the rules about module names and naming conventions will be discussed in this chapter, the significance of a module name as an
indicator of relative module quality is discussed later.

Module names are an important part of this notation. The rule is that the name
of a module not only should summarize what that module does but should take a

TABLE 9.2.1-1:   Examples of Modules in Different Programming Languages

Programming Language	Example
ADA*	Function
	Procedure
	Task
BAL	Csect
"C"	Function
COBOL	Program
	Paragraph
FORTRAN	Subroutine
	Function

*ADA is a trademark of the U.S. Department of Defense.

TABLE 9.2.1-2:   Examples and Counterexamples of Module Naming Conventions

Potentially Acceptable Verbs	Usually Unacceptable Verbs	Potentially Acceptable Nouns	Usually Unacceptable Nouns
Validate	Process	Target	Information
Edit	Input	Client_Name	Data
Update	Output	Account_Number	Input
Compute	Confirm	Trajectory	Output
Determine		Track	
Calculate		Identity	
		Current_Balance	

certain form as well. Similar to our naming conventions for processes in dataflow diagrams, module names each consist of a strong verb and an object noun. The names used in the Structured Design phase are taken directly from the names of the corresponding processes in the dataflow diagram from which the initial Structured Design was derived. Although *all* names have a potential for abuse, some examples and counterexamples of acceptable and unacceptable names are presented in Table 9.2.1-2.

A rectangle or a square is used to indicate a module. The name of the module, utilizing the naming conventions described above, is contained within the rectangle (Figure 9.2.1-1).

9.2.2 The Call

A module can perform its designated task only when it is in control. A CALL, a PERFORM, or some other such mechanism can be used to transfer control from one module to another. That module may then, as part of its task, call another module, and so on. A protocol is associated with the invocation of other modules. It is that each module eventually returns control to the module that called it, and so on until control reaches the module which is at the top of the hierarchy. That module may do any of several things, including returning control to the system. The notation for the CALL is demonstrated in Figure 9.2.2-1. Note that the conditions (if

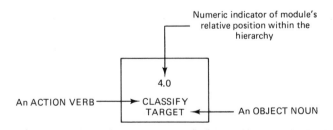

Figure 9.2.1-1:   Details of the Module Symbol

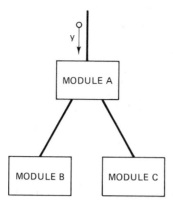

This Structure Chart means that the pseudocode or code which defines what
Module A does contains <u>at least</u> one statement like CALL MODULE_B and
<u>at least</u> one statement like CALL MODULE_C. It is possible that Module A
may use the value of data item y to decide which module to call. However,
we cannot tell this from the Structure Chart alone.

Figure 9.2.2-1:   An Example of the CALL Notation

any) under which a particular module actually does CALL another are not present
on the Structure Chart. Also, at least in the basic notation, it is not apparent from the
chart whether or not a module calls another more than once. The Structure Chart
indicates only what *can* happen with respect to CALLS, not necessarily what *will*
happen when the software executes. This is dependent on the nature of the data
processed and interrupts.

9.2.3 The Data Couple and the Control Couple

As mentioned earlier, two types of information can be communicated between
modules—control and data. As there is often a great deal of confusion (even among
software engineering "experts") regarding the difference between these two types
of information, some explanation is in order before we discuss the notation.

The information which passes in either direction (i.e., from or to the calling or
called module) can take on either of two characteristics. In one case, the information is strictly inherent in the problem being solved. One might call it information
that is an *intrinsic* part of a description of the enterprise [2]. In the case of a system
to assist in billing a dentist's patients, examples of such data would include the
patient's name, address, type of service performed by the dentist, and fee. This
enterprise-inherent data is graphically depicted on the Structure Chart when it is
transmitted between modules. Since it refers to information which is *data* oriented,
individual members of this class of information are referred to as *data couples*. A
specific type of symbol is used to depict data couples on Structure Charts.

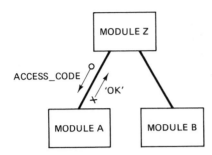

Figure 9.2.3-1: Simple Example of the Use of Couple Notation

The second type of information depicted on Structure Charts is not inherent in the nature of the enterprise. It was not present during the analysis phase but exists solely to support the design and subsequent phases. This type of information exists as a direct result of the fact that this system is or is going to be automated. Examples for the dentist system referred to above include an end-of-file flag, a flag to indicate that no more space is available on the disk, a flag to notify a module that the input received from the operator was unacceptable, and other such status-oriented information. Note that the dentist does not have or use such flags now in the manual system. These indicators of condition exist because we wish to automate the system and, as a result, flags will be required. This type of information is referred to as control oriented. Individual instances of one module transmitting one (or more) control indicators of this type are referred to as *control couples*. A notation is used that is somewhat similar to that employed to indicate data couples.

Examples of both types of couples are presented in Figure 9.2.3-1. The main point to keep in mind, in determining whether or not a particular piece of information is of one type or another, is whether or not it is an inherent part of the enterprise which this system is intended to support. If it is, it is a data couple. If it is strictly a result of implementation, then it is a control couple.

9.2.4 Alternative Structure Chart Notational Approaches

In the original work of Stevens, Myers, and Constantine on Structured Design in 1974 [3], the issue of system size was not emphasized. Today, systems with Logical Designs containing hundreds or thousands of modules are common. When one considers the degree to which such systems will fan out, it is easy to see that a very confused-looking diagram can result. But fan-out is not the only contributor to confusion. There are several others in the original form of the structure chart:

- An arrowhead is often used redundantly at the lower end of the CALL symbol. The fact that one module is physically lower on the chart than another means *explicitly* that the higher module CALLS the lower one.

- The fact that more than one module may call another leads to several sets of crossed CALL lines. This can be very confusing and frustrating.

- The angular placement of CALL lines makes *simple* automated support difficult.

- The numerous named or labeled DATA and CONTROL couples tend to clutter the diagram. What is worse, users of the structured methods often employ nearly unintelligible abbreviations to reduce the character density.

- The DATA and the CONTROL couples *both* have the same shaped end—a circle. One is filled in and the other empty. Some designers, users, and reviewers confuse the two.

Faced with all of these shortcomings, some may be tempted to do what the author of at least one popular text did—eliminate couples altogether. However, this would greatly reduce the utility of the Structure Chart. For example, being able to visually spot some data or control item racing around the chart does enhance our ability to refine the design. Fortunately, in the course of many project consulting engagements, several effective and less stringent schemes have been developed which retain the essence of the Structure Chart while overcoming its large-system problems. Although we will examine these schemes one at a time, they can be and have been used successfully in combination. Remember, these are not *proposed* alternatives but ones which have been used on *real* projects. They may or may not be acceptable to a given project organization. It is recommended that the needs and preferences of your project be used to select the approach most suitable. Also, do not be afraid to change the approach during the project if it is not meeting the need. (Note: We have left out the couples on the examples below for ease of explanation.)

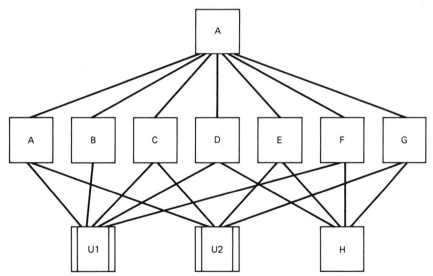

Figure 9.2.4.1-1:   Example of Crossed-Lines Problem in Structure Charts

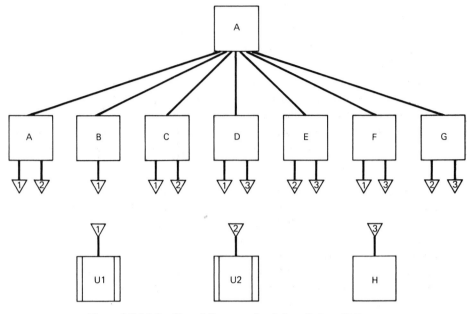

Figure 9.2.4.1-2:   Use of Connector Symbols to Reduce Clutter

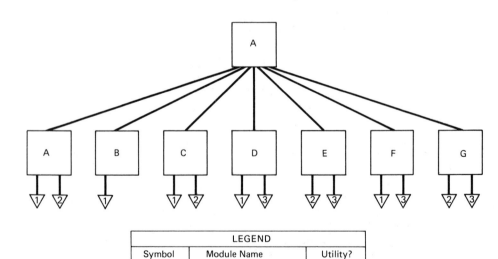

LEGEND		
Symbol	Module Name	Utility?
1	VALIDATE_CODE	yes
2	LOCK_FILE	yes
3	GET_TEMP_SPACE	no

Figure 9.2.4.1-3:   Use of Connector Symbols and Supplementary Table

9.2.4.1 The "cross not this line" variation

Popular modules and utilities can often lead to an exasperating display of crossed lines (Figure 9.2.4.1-1). Two related but somewhat different schemes have been used to address this problem:

Use of connector symbols. Some have found it useful to use connector symbols to reduce clutter (Figure 9.2.4.1-2).

Use of connectors and tables. Often, there are too many low-level modules to make the connector approach practical. This problem is aggravated if the Structure Chart has many levels to it. This approach employs a reference table on *all* Structure Charts for the reader to refer to for explanation (Figure 9.2.4.1-3).

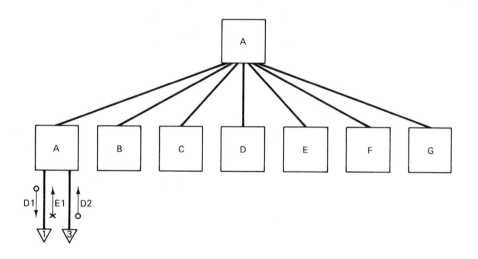

LEGEND		
Symbol	Module Name	Utility?
1	VALIDATE_CODE	yes
2	LOCK_FILE	yes
3	GET_TEMP_SPACE	no
Symbol	Item Name	Type
D1	CODE	DATA
D2	TEMP_SPACE_ADDR	DATA
C1	VALID_FLAG	CONTROL

Figure 9.2.4.2-1:   Example of the Extended-Symbol-Table Approach

9.2.4.2 The "players and scorecard" (extended symbol-table) variation

In this approach the use of a table to explain symbols is extended to its logical limit. It addresses several problems at once. What has increasingly caused problems in the use of Structured Design notation is the increased interest in and use of the dictionary. Better, more elaborate, long data and control item names have made the use of symbol labels clumsy. In our previous examples, we purposely left out the use of data and control couples. If we had included them, the clutter might be unmanageable. The extended symbol-table approach enables us to use labels that are convenient without clutter (Figure 9.2.4.2-1).

9.2.4.3 A compact hybrid

On one project people got so harried that the earlier approaches did not reduce the clutter enough. As a result, they devised a more compact scheme (Figure 9.2.4.3-1). In it, a set of notational conventions are used in conjunction with the symbol-table concept described earlier. Note that this scheme eliminates the sometimes objectionable couple symbol. It also does away with the use of offpage connectors, since each segment of the Structure Chart contains the input(s) and output(s) associated with the boss module.

9.2.4.4 A slightly radical departure

The degree of fan-out which occurs in the standard Structure Chart is indicated by the often steeply angled lines emanating from one or more boss modules. This mechanism documents the "span of control" of that boss. The CALL symbol (with or without the redundant arrowhead), the data couple, and the control couples are all integral parts of the Structure Chart notation. Hence, the fundamental elements of the Structure Chart are:

Control hierarchy
Span of control
Information communication (data or control)
Module population

The authors of the Structure Chart, in its original form, probably did not *ever* intend it to be used on systems as large as those being developed today *and* be implemented exactly as they had drafted it by hand. Angular lines and circles are incompatible with simple, inexpensive output devices. Those who are automating the use of the Structure Chart have, for the most part, ignored a fundamental rule of automation:

NEVER emulate a manual process on a computer!

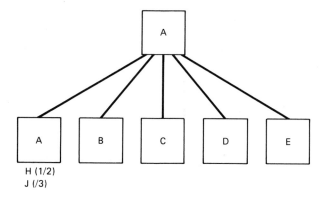

H (1/2)
J (/3)

LEGEND		
Symbol	Module Name	Utility?
H	VALIDATE_CODE	yes
I	LOCK_FILE	no
J	GET_TEMP_SPACE	no
Symbol	Dictionary Item	Type
1	CODE	DATA
2	VALID_FLAG	CONTROL
3	TEMP_SPACE_ADDR	DATA

Notation Conventions:

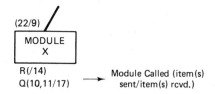

Figure 9.2.4.3-1: Example of the Compact Hybrid Approach

This oversight is unfortunate, because most of the resource and effort on such tools has been expended on a graphics, *not* an analytic, capability. The automators failed to realize that the results of design are much more than pictures. Without the analytic support (e.g., dictionary), the pictures are worthless.

The admonition above can be easily followed if one takes advantage of what the computer has to offer while retaining the *essence* of what we are trying to communicate. Figure 9.2.4.4-1 depicts a notational scheme which is easy to automate *and* retains the essence of the Structure Chart.

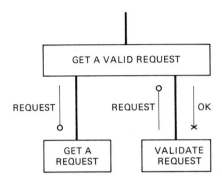

Figure 9.2.4.4-1:   Structure Chart Notation
Designed for Automation

9.3 SYNCHRONIZATION OF PROCESSES IN REAL-TIME AND MULTIPROCESSOR SYSTEMS

Since Structured Design was first published [3], the nature of the software design process has changed dramatically. In the early seventies, attention focused on the transformation of manual tasks into automated systems. There was a heavy emphasis on the definition and development of batch applications as well as some real-time, interactive ones. At that time, Structured Design and its Structure Chart notation were more than adequate to the task. Today, however, the emphasis is on taking those batch applications with fairly well defined and understood requirements and replacing them with interactive systems, database systems, and others which are more capable of dealing with an environment which is changing rapidly. The original form of Structured Design needs to be extended to accommodate these new requirements.

Two primary areas of concern with the Structure Chart notation involve (1) information hiders and (2) processes which are running or executing in the same *logical* time. We refer to *logical* time because the Heisenberg Uncertainty Principle [4] clearly indicates that the processes cannot execute at *exactly* the same time. What notation can we use to indicate such situations? Information hiding will be discussed in Chapter 14. The synchronization of cooperative and multiply occurring modules is our next task.

In Structured Design in its original form, the coordination of modules occurs through the use of flags, calls, and returns. These all involve a boss module. For example, in Figure 9.3-1, Module X calls (or may call) Modules A and B. Only after MODULE A has completed its task successfully (as indicated by the flag) does it call MODULE B. This works out well in batch and other applications where the execution time is less critical than in real-time systems. In a real-time system, it is often the case that MODULE A and MODULE B are both executing in the *same logical time*. Even so, we could use the same signaling arrangement. We need not execute them in the same logical time, but the alternative could be very "expensive" with resepct to execution time. The instructions which the system would execute before the actual transfer of control and the return of control would account

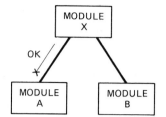

Figure 9.3-1:   Simple Example of Structured Design's Original Approach to Module Coordination

for this "time expense." When they do execute at the same logical time, we take advantage of machine capabilities by changing what was a sequential situation into a parallel one.

Another situation which can be "expensive" involves the synchronization of modules. For example, let us say that MODULE A should not proceed beyond a certain point in its execution if MODULE B has not proceeded beyond a certain point in its execution. In other words, while the execution of the two modules is separate, it is not entirely independent. This is a fairly common condition when one module is updating files that another needs to use. The actions of the two modules must be synchronized in some way. To do this, we can SUSPEND the execution of MODULE A at the appropriate point if and only if MODULE B had *not* completed its execution beyond a certain point appropriate to it. One way to handle this would be to split MODULE A into two modules (say, MODULE A and MODULE AA). This is shown in Figure 9.3-2. This approach, however, is time consuming and reduces the effective capabilities of the machine. That is, we lose out on possibilities for having multiple processes executing in the same logical time.

A better way to accomplish synchronization is to use a protocol of SUSPEND and RESUME commands. This would allow each module to signal its status to the other *without* significantly reducing processing throughout, destroying parallel processing capabilities, or invoking the boss module. Here is how it would work. Again, MODULE A must not proceed beyond a certain point in its processing if MODULE B has not completed the first phase of its processing. To accomplish this, MODULE A has in it a statement which will cause it to suspend its processing at that point if and only if MODULE B has *not* completed its first phase of processing.

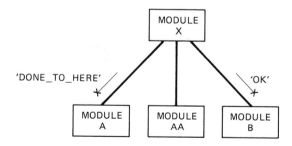

Figure 9.3-2:   A Conventional Means of Synchronizing Module Activities

This is the same situation we had before—with one important exception. We would like to accomplish the signaling of the condition of the processing performed by Module B *without* incurring the overhead involved in a series of CALLS and RETURNS to some boss module. This is depicted graphically in Figure 9.3-3.

The notation in Figure 9.3-3 indicates that MODULE A will SUSPEND itself until MODULE B posts a RESUME to the syncbox (indicated by the S-ended rectangle). If MODULE B posts a RESUME to the syncbox *before* MODULE A reaches its SUSPEND point, MODULE A will *not* suspend but just keep on going. The number in the syncbox is its unique identification. Each syncbox will have its own number. The aim is to reduce any confusion that may be caused by having a single module which requires more than one syncbox (either SUSPEND or RE-SUME). Also, this numbering enables us to describe any assumptions or details about the synchronization for each of the syncboxes in the Lifecycle Dictionary. Any module may be related to any number of syncboxes. Also, two modules or more may have several different SUSPEND and RESUME points with respect to each other. This can be indicated through the use of several syncboxes involving only those two modules. Also, a module may require that more than one other module has executed beyond some phase of its processing. Such situations can be indicated through the use of more than one syncbox. The module with such a requirement would have more than one SUSPEND, possibly in succession, in its pseudocode.

A generic description of the notation is presented in Figure 9.3-4. Note that where there will be multiple copies of the module(s) involved, this is indicated by means of a numbered diamond. This notation is necessary when we have, for example, two processors, each with its own copy of one or more of the modules involved, as demonstrated in Figure 9.3-5.

Regardless of whether or not there is a need to have copies of these processors, every instance of a SUSPEND should be accompanied by a time limit. In

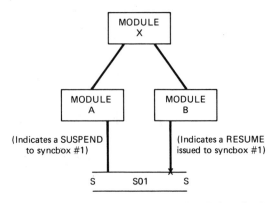

Figure 9.3-3:   Use of the Syncbox Approach to Reduce Overhead

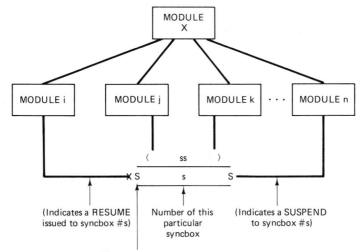

[ss indicates the number of copies of this syncbox present]

[⟨ss⟩ indicates the total number of copies of this syncbox present]

Figure 9.3-4:   Generic Description of Syncbox Graphics

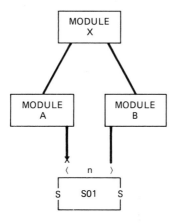

where n = the total number of copies of the syncbox. If there is only a
single copy of the syncbox, the ⟨ n ⟩ symbol is not shown.

S01 = the number of this particular syncbox used to reference it in
the dictionary and on the Structure Chart

Figure 9.3-5:   Graphically Describing Multiple Processing Situations

this way, if a process has suspended itself dependent upon the completion of another process, we can detect if this other process will have experienced some difficulty. That is, it could be hung in some sort of loop. In such an instance, the module that suspended itself might never execute again. The use of a timer enables us to protect the system against this. If the timer expires, then we know that a problem exists, and some documentary or remedial action can be taken, possibly without the need to stop processing altogether.

9.4 MESSAGE PASSING BETWEEN REAL-TIME PROCESSES

A concept similar in nature to the syncbox is that of the mailbox. Mailboxes are used to transfer data, *not* flags. In this case, the problem we are attempting to overcome is a common one in real-time systems development. It addresses three issues:

> **Contention resolution.** No matter how many tasks are waiting on a message, only *one* receives it.
>
> **Handshaking.** This replaces the function taken for granted in a Procedure call.
>
> **Timing independence.** The receiver of a message may ask for data before it is sent and still get it.

The notation for mailboxes is shown in Figure 9.4-1. The generic form of the mailbox notation is presented in Figure 9.4-2.

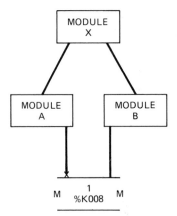

Figure 9.4-1:   Example of Mailbox Notation

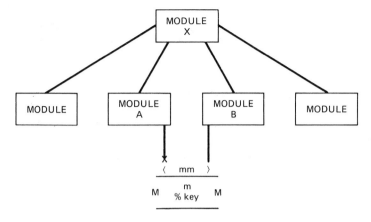

where mm = the total number of copies of this mailbox. If there is only
    a single copy of the mailbox, the ⟨ mm ⟩ symbol is not
    shown.

    m = the number of this particular syncbox used to reference it in
    the dictionary and in the Structure Chart

% key = identifier of lock/unlock parameter for dictionary reference
    with restricted access to values

Figure 9.4-2:   Generic Description of Mailbox Graphics

9.5 STANDARDS AND GUIDELINES FOR USE WITH STRUCTURE CHARTS

The most successful use of Structure Charts occurs when a set of operating rules are adhered to. A suggested set of these is presented in Figure 9.5-1. These are both standards and guidelines, in that they are both advised and may be made into hard-and-fast rules. The reader is advised to consider which, if any, of these may be safely disregarded.

The symbol for a module is a rectangle.

Each module must have a label.

Module labels are composed of a transitive (or action) verb and an object noun.

With the exception of utility modules and the system-level boss module, each module must have a numerical identifier unique within that system of modules.

Module numerical identifiers are composed of a local numerical identifier concatenated with the numerical identifier of its identified boss module.

If a module is called by more than one other module, and it is <u>not</u> a utility module, the module which calls it the most is its boss.

Figure 9.5-1:   Standards and Guidelines for Use with Structure Charts

Modules may communicate information only via CALLS or information hiders.

A module which is communicating information via a CALL may only communicate it directly (i.e., not through an intermediary) to the module it intends it will use the information.

A module may not call or invoke another module which is higher in the hierarchy than it is.

Modules which are utilities (i.e., are available as part of the environment and do not need to be developed) are depicted symbolically with double vertical lines on the left and right end of the rectangle.

Although a module may be called by more than one other module, each module may have one and only one identified boss.

No two modules, data couples, or control couples may have the same name.

No data couple may have the same name as a control couple.

The operation which takes place in a module must be described with pseudocode.

All data couples, control couples, pseudocode, syncboxes, mailboxes, and data item names must be defined in the dictionary.

Synchronization of modules in multiprocessing and other real-time systems may be accomplished through the use of a SUSPEND–RESUME protocol employing syncboxes or mailboxes, as appropriate.

Mailboxes and syncboxes are not actually modules but are shown graphically for explanatory purposes.

Mailboxes and syncboxes have a unique number.

Syncboxes may be employed by two or more modules for synchronization.

Mailboxes may be employed by two or more modules for communication and synchronization.

If multiple instantiations of modules and their syncbox(s) are to be employed, the number of instantiations is indicated between a pair of angular brackets (greater and less than symbols) immediately above the syncbox symbol.

The syncbox symbol is a narrow rectangle with an S at each of its short ends.

The mailbox symbol is a narrow rectangle with two M's at each of its short ends and a symbol indicating the number of occurrences and the identifier of the mailbox key preceded by a percent sign (i.e., m % key).

Figure 9.5-1:   (continued)

9.6. RELATIONSHIP OF STRUCTURE CHARTS TO OTHER TOOLS IN STRUCTURED DESIGN

The Structure Chart is one of the three basic tools of Structured Design. The other two, the Design Data Dictionary and pseudocode, both support and are supported by the Structure Chart (Figure 9.6-1). Note that all of the data described on the Structure Chart via data couples and control couples must be defined in the Design Data Dictionary. Similarly, the pseudocode which describes the operation of each module was derived from the structured English for each Structured Analysis process that became a module. There will not normally be a one-to-one correspondence between process bubbles in the analysis and modules in the design. Some processes

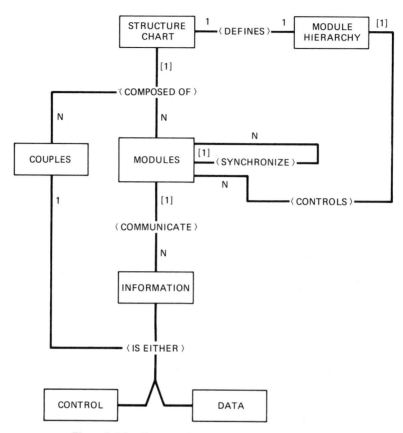

Figure 9.6-1:   Structure Charts and Their Components [S1]

may get broken up into multiple modules, some groups of processes may be collapsed to form single modules, and modules will have to be created to accommodate the need for error modules, reporting modules, and data handling modules.

Just as in the case of Structured Analysis, the tools in Structured Design form an integrated set. The successful use of the Structure Chart, as one of these, depends on the correct and disciplined use of the other tools. As can be seen from Figure 9.6-1, Structure Charts are not very useful without the existence of the data dictionary defining each data item included in the system, pseudocode for each module describing how it accomplishes its task(s), and an Entity-Relationship diagram describing the nature of the database environment within which this system will be run. Table 9.6-1 shows a list of questions that are often asked about the information found on the Structure Chart, together with the appropriate tools which provide the answer(s).

TABLE 9.6-1: Questions Frequently Asked About Structure Charts

Do Structure Charts show the sequence of execution of the system?

No! Although most software designers will draft the Structure Chart so that what would normally occur first in the operation of the system appears on the left and what would occur last appears on the right, the Structure Chart is not intended to show the order in which modules gain or relinquish control.

If one module is shown calling another module on the Structure Chart, does this mean that this must occur at some time during the execution of the system?

No. The structure Chart CALL symbol only documents the fact that a given module contains one or more statements which will cause it to pass control to another module. When the system is in operation, the data being processed may be such that these CALL statements are not actually executed.

Must all modules in the system be shown on a single Structure Chart?

Only if it is convenient and it would not cause confusion. Most systems contain so many modules that it is physically impossible to create a single diagram which depicts them all in an understandable way. In such cases, it is better to segment the Structure Chart, using off-page connectors. A reasonable segmentation rule is to show the top or boss module and its immediate subordinates on a single diagram. Each of its subordinates would be shown as a boss on another, together with its immediate subordinates. If any of these subordinates also have subordinates, this process can be reapplied until the lowest-level modules are described.

In order to simplify the diagram, can Structure Charts be drawn which do not show the DATA and CONTROL COUPLES?

This is a common question and involves a conflict in terminology. A Structure Chart, by its very nature and intent, shows the population of modules, their inherent control hierarchy, *and* the information that they communicate. If one or more of these are left out, we are not dealing with a Structure Chart any more.

9.7 SOME ALTERNATIVE FORMS OF STRUCTURE CHART NOTATION

As one might guess, the Structure Chart is such a useful tool that some software engineers just cannot resist the temptation to modify it. Certainly, we cannot hope to present all the variations that have been used. Instead, we will present some which have proven most useful.

Four types of information normally excluded from Structure Charts leave some software engineers clamoring for more:

Execution speed

Module size

Status of software

Potential number of executions of a module

Some software engineers have found a way to indicate the potential number of times a module will be called right on the Structure Chart. Figure 9.7-1 depicts a common way to include this type of information. Note that where limits are known, they are included. Modules which possess the potential for becoming "black holes" for execution time (e.g., module E in Figure 9.7-1), are easy to spot. Such information early on will aid management in assigning various parts of the system

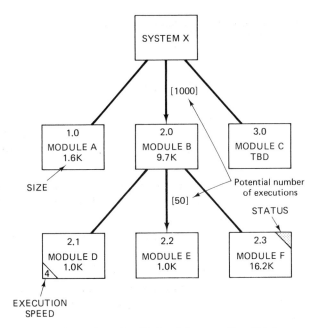

Figure 9.7-1:   Example of Tailored Structure Chart Notation

for implementation as well as formulating a scheme for the order in which the implementation should occur.

The potential size or estimated size of a module is a concern to many. Where the notation has been changed, the approach shown in Figure 9.7-1 is typical. The source of a required modular function has been shown using either of the approaches shown in Figure 9.7-1.

Given a little time and imagination, most software engineers could conceive of some notational scheme or another that meets a specific need they may have. But where the variations we have related or some new ones are employed, we should bear in mind that the more we ''clutter'' the Structure Chart, the less effectively it will relate the most important class of issues we can consider—architectural ones.

9.8 STRUCTURE CHARTS—A SUMMARY

The Structure Chart is a graphic representation of key aspects of the software system design. It is an important part of Structured Design. As we will demonstrate in subsequent chapters, it is not the only tool of Structured Design. However, it does provide a concise presentation of the use of the other tools of Structured Design.

The Structure Chart may take many forms. It is flexible enough to be adapted to a wide variety of situations. We have demonstrated the use of a few modifications. Many more are possible. The number of alternate forms of Structure Chart is limited only by one's imagination.

REFERENCES

1. D. Couger and R. A. Zawacki, "Key Factors For Motivating Computer Professionals," in *Systems Analysis and Design: A Foundation for the 1980's,* ed. W. W. Cotterman, et al. New York: North-Holland Elsevier, 1981, pp. 417–28.

2. P. Chen, "The Entity-Relationship Approach to Logical Data Base Design," The Q.E.D. Monograph Series on Data Base Management, No. 6, Wellesley, Mass., Q.E.D. Information Sciences, Inc., 1977.

3. W. P. Stevens, G. J. Myers, and L. L. Constantine, "Structured Design," *IBM Systems Journal,* Vol. 13, No. 2 (May 1974), pp. 115–39.

4. Z. Heisenberg, Physik, Vol. 43, p. 172, 1927.

S1. Extracted from the seminar, "Structured Design of Real-Time Systems Seminar," by Software Consultants International, Ltd.; Kent, Washington; Copyright 1985, 1986. Reprinted with permission.

CHAPTER 10

Making the Transition
from Structured Analysis

Structured Analysis helped us to define the partitioning of the problem in a systematic way. We were also able to document the information used by the system; the modes in which the system may be expected to operate; the information, relationships, and policies about that information; and the rules and policies which govern the system's operation. The partitioning was accompanied by a description of the levels of decomposition that were appropriate to the system. In developing these levels, we used the analogy of parent-child or archetype-surrogate relationships. With respect to software design considerations, the parent of a given set of processes controls those processes. When one considers what the "boss" module of a given set of modules does, it plays the role of parent. Hence, in making this transformation, we will be creating an architecture in which the parent processes become boss modules. If some of the subordinate processes are themselves parents, then their subordinates will be lower on the module hierarchy. This translation process is just to get us started. It will result in a design architecture which we will improve upon.

In Structured Design, we utilize the partitioning set forth during Structured Analysis. We do this by proceeding through a transition process. This transition is largely a conversion process. However, the development of a viable Structured Design, one which we would wish to implement, is anything but a conversion process. It involves the application of a variety of quality criteria and other techniques which we shall discuss later.

Many people who promote the use of the Structured Methods leave the impression that this transition is an easy and straightforward one. This is far from the truth. True, the first cut of the design is relatively easy to obtain when a quality analysis has been done, but things get "fuzzy" very quickly after that. "Fuzzy" means that some of the models that were built may undergo some important changes as we proceed closer and closer to a design that we could call a system blueprint. Two areas of particular concern are related to real-time, parallel-processing systems and systems which involve interaction with a database of moderate to large size. In both of these cases, the respective models that were developed during analysis will be modified extensively, perhaps beyond recognition. This does not mean that an inadequate job was done during the analysis, but it does mean that the goals and aspirations of the software engineer are very different during design from what they were during analysis. During analysis, we were attempting to define the problem. That allowed us to deal with an "abstract" space in which the realities of timing, sizing, hardware limitations, hardware architecture, and so on did not apply. This flavor of abstraction is retained during the initial design activity until we are sure that everything that was developed during the analysis has somehow found its way into the logical (or abstract) design. Once this is accomplished, we begin to introduce the issues that must be addressed in order to obtain a realistic compromise between what is ideal and what is possible within the constraints imposed upon us by the real operating environment.

This chapter describes and provides examples of how the leveled dataflow diagrams which were developed during the analysis phase may be converted into a *first-cut* design. We refer to this as a first-cut design because it is just that—that is, a starting point for the design activity, not an end or final product. Even after it has been refined, this form of the design will not be suitable for implementation. It will constitute the Logical Design.

10.1 CONTENT OF THE ANALYSIS PACKAGE

Before we discuss the translation process, we need to make sure that we have a complete, consistent, and correct Structured Analysis. Figure 10.1-1 presents the information that is present in the package and where it is going with respect to project phase. The set of information that we need we shall refer to as the *analysis package*. This package is more than just a set of dataflow diagrams. Certainly, these diagrams are necessary, but they are *not* sufficient to provide what we need to begin the design activity. The *minimum* set of information that we will need is listed below:

- A set of leveled, balanced dataflow diagrams
- A dictionary which defines *all* dataflows down to the element level
- Pseudocode describing the operation of each process in each dataflow diagram

	DEVELOPMENT PHASE		
TYPE OF MODEL	STRUCTURED ANALYSIS	STRUCTURED DESIGN	IMPLEMENTATION
PROCESS	Dataflow Diagrams Dictionary[1] Pseudocode	Structure Chart Dictionary[2] Pseudocode[4]	Code Dictionary[3]
INFORMATION	Entity-Relationship Diagram Dictionary[1]	Database Design Dictionary[2]	Implemented Database Dictionary[3]
EVENT	Event Model[5] Dictionary[1]	Event Model[6] Dictionary[2]	Queuing Model Dictionary[3]

[1] This dictionary includes data definitions: pseudocode for each process in the analysis model; descriptions, attributes, and content for the entities and relationships in the E–R diagrams; and descriptions of the events and states in the Event Model.

[2] This dictionary contains everything found in the analysis version of the dictionary plus definitions of all flags and pseudocode for all modules that were added as part of the design process or not otherwise present in the analysis.

[3] This dictionary has been updated to incorporate all information developed during the design activity which has been modified as a result of implementation considerations.

[4] The pseudocode referred to here incorporates any and all flags employed in the Structure Chart. All pseudocode is contained in the dictionary as well. Much of this may be identical to or based on the pseudocode that was developed during the analysis.

[5] Event Model refers to a simplified form of state transition diagram.

[6] This Event Model would be further refined and leveled based on design "discoveries."

Figure 10.1-1:   Information Used by Project Phase [S1]

- An Entity-Relationship diagram for the system
- Dictionary entries which describe each entity and relationship, their attributes, purpose, content
- An Event Model consisting of the events and states the system may operate in
- Dictionary entries which define each state and describe the events and their logical relationships
- Verification that the analysis has addressed all requirements

Given that we have the above information in a convenient and consistent form, we may begin the process of converting the analysis into a starting point for design activity.

10.2 TYPES OF SYSTEMS

The nature of the systems with which we will be dealing affects the structure and organization of the design. In this regard, it is useful to note that there are two major classes of systems: (1) those in which the same input will always produce the same output, and (2) those in which the same input may or may not produce the same output.

The first case described above is commonly associated with "batch"-type systems. That is, the user inputs the data, initiates the system, and the result is always the same, provided the input data values are the same.

The second case refers to a type of system that has many different modes of operation. The system response in the first case is determined strictly by the value of the information fed into it. But the second case presents a different problem. The system has different modes of operation (also known as states). The information which is valid in one mode of operation is not necessarily valid when in another mode of operation. Hence, the response of the second type of system is a function not just of the single input which the user enters but, rather, of the series or sequence of inputs from the user and external sources which can cause the system to change its mode of behavior.

This second type of system causes software designers the most difficulty. It is referred to by some as a "transaction-oriented" system. The first case is a special case or degenerate case of the second. It is referred to by some as a "transform-oriented" system. It is a degenerate case in that it operates in a single (rather than multiple) state. As we shall see later in this chapter, transaction-oriented systems are really a combination of transform-oriented subsystems under the control of one or more "boss" modules.

10.3 RECOGNIZING TYPES OF SOFTWARE SYSTEMS

The dataflow diagrams which describe a particular system often clearly indicate what type of system we are dealing with. Certainly, we may know long before that, but recognizing which part of the system is one type and which is another will aid our ability to produce a quality result with minimum effort.

As in the case of dataflow diagram development, the data holds an important key. When we examine the input to our system or a portion of the system, we need to determine whether or not the system will require the data to have a "tag" or indicator to tell it what type of data is going to be processed. The system has no way of knowing a priori what data will be received or in what order. For example, a disk operating system may enable a user to change the name of a file by entering:

```
RENAME OLD_FILE_NAME.OLD_EXT NEW_FILE_NAME.NEW_EXT
where
    OLD_FILE_NAME is the current name of the file
    OLD_EXT is the extension of the current file
    NEW_FILE_NAME is the new name of the file
    NEW_EXT is the extension of the new file
```

To the system, the data received from this command "looks" like:

```
| OPERATOR  |  OPERAND_1  |  OPERAND_2 |
```

←''TAG''→   ←''COMMAND CONTENT''→

Note that the information received from the user was composed of two distinct parts. The first was a TAG to tell the software what kind of processing would be required, and the second was the actual COMMAND CONTENT which the software would operate on.

Today, most systems of interest involve interaction between user and software or at least some sort of interrupt capability. Interrupts are similar to the TAG in that the type of processing being requested is identified by means of the interrupt, while the information to be processed can be identified by a variety of protocol mechanisms. The fact that these transaction-oriented systems may be required to provide several different types of processing, any of which can be triggered via the appropriate "tag," is generally easy to detect on the dataflow diagram. If the dataflow diagram has one or more of its primary input dataflows "fanning out," then it is highly likely that we are dealing with a transaction-oriented system.

An example of what is meant by fan-out is shown in Figure 10.3-1. In it, the actual values for the reservation request tag are indicated on the diagram within single quote marks. The fan-out occurs where the three dataflows (of sorts) diverge. The three dataflow lines are such that any given request that comes down the pipeline will proceed down one and only one of these three paths. The label that is provided identifies which path will get the information. Owing to its impact in the design phase, it is best to think of these paths as actually transporting no data. They are only a convenient way of indicating which of the processes on the diagram will be "turned on" or operable under the condition stated (e.g., if the request is a CONFIRM, that process will be "on").

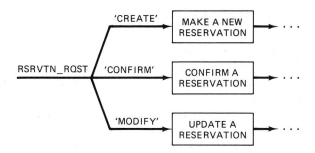

(Note: Details of what happens to the data have been omitted)

Figure 10.3-1:   An Example of Dataflow Diagram Fan-Out

Does this approach violate one of the rules set forth regarding the incorporation of control information (i.e., flags) into dataflow diagrams [1]? Perhaps, but not exactly. The admonition to avoid the incorporation of control information was intended to prevent the creation of dataflow diagrams that are, in reality, flowcharts. Not that flowcharts are inherently useless, but we were trying to perform a different kind of analysis than flowcharts support.

In this case, we are using this specialized notation to indicate which process will actually be using the data. The difference may be a subtle one, but at the time of publication of the original text in this area, the experience base with this method on real-time, interactive systems was limited, and the topic was not addressed as heavily as it might be. The use of the 'tag' indicator as shown above seems reasonable if for no other reason than that it works and avoids communication of misinformation. It is incorrect and detrimental to the design to indicate that the request goes to all three processes, because it does not.

A supplemental approach to inspecting the dataflow diagram is to inspect the dictionary definition of the dataflows coming from the user or primary data source. If we find a definition such as

```
RSRVTN_RQST = TYPE_INDICATOR + OPERATOR_ID
```

then we can be reasonably sure that the system we are dealing with is transaction oriented.

In reality, transaction-oriented systems are a mixture of transform and transaction centers. That is, the transaction centers are not the only things which populate the Structure Chart. So our concern should center less on what type of system we are dealing with and more on the details surrounding the content of the analysis. Our objective should be to ensure that *everything* that was contained in the analysis finds its way into the first-cut design—nothing more and nothing less.

10.4 A BOSS, A BOSS—MY KINGDOM FOR A BOSS!

Some authors spend a great deal of time and space on the issue of picking the "right" boss for the system. It is implied that one must be very careful in selecting which process(es) will be a boss and which will be subordinate. Experience shows that this emphasis has a negative effect on some analysts and designers. They are so concerned about "getting it right" that they are reluctant to make a decision. In this section we will demonstrate the use of two simple techniques which enable one to quickly make the transition from the analysis to design packages. The importance of the initial structure is often overstated, since this first-cut design will be restructured several times with many changes in the hierarchy.

10.5 Problems with Deriving Designs from Structured Analyses

Although the Structure Chart is the most prominent part of Structured Design and usually receives the most attention in discussions about deriving Structured Designs, we will take a more broadbased, systems approach and discuss the derivation of all the elements of Structured Design. Figure 10.1-1 shows the relationship between the tools of Structured Analysis and those of Structured Design. Note that each tool in Structured Analysis is utilized by Structured Design. The first-cut design is actually derived from the analysis. A useful side effect of such a scenario is that the bond between the specification (i.e., the analysis) and the solution model (i.e., the design) is strengthened by the use of Structured Analysis with Structured Design. As indicated in Figure 10.1-1, the Structure Chart that we obtain from the dataflow diagram is a *first cut*. It is *not* what would be considered a finished or final blueprint. Its shortcomings include:

> **Serious lack of error processing.** As discussed earlier, Structured Analysis downplays error-processing issues and accentuates the normal processing.
>
> **Input/output processing.** The reporting and data-gathering processes were downplayed in Structured Analysis.
>
> **Validation and editing.** These were minimized.

All of these cannot continue to be ignored or downplayed during design. They must be addressed, together with further refinement of the existing partitioning, modification to accommodate design quality considerations, and compromising to respond to implementation environment characteristics.

10.6 DERIVATION PROCESS

This technique may be popular because it requires almost no thinking. It involves emulating what is contained in the analysis hierarchy by means of a simple translation process. Figure 10.6-1 depicts a set of leveled dataflow diagrams. We assume that the supporting information that complements those diagrams is also present (i.e., dictionary entries, pseudocode, balancing). Notice that the topmost diagram, the Context level, contains a single process. The topmost process will become our "top boss." The processes which appear at level '0' will report directly to the top boss. Each of the immediate subordinates to the level-'0' processes (now modules) will be modules reporting to their respective parents. We continue this procedure until all the processes described on the leveled dataflow diagrams have been represented in some way on the Structure Chart.

It is recommended that the information flow associated with each level be

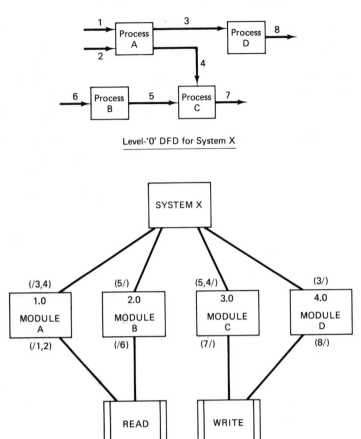

Level-'0' DFD for System X

Equivalent First-Cut Structure Chart

Figure 10.6-1:   Translation of a Level-'0' DFD into a First-Cut Structure Chart

resolved prior to proceeding to a subsequent level. Note that the only means by which we will allow any child or subordinate module to communicate information to any other module will be through the parent module. This may seem somewhat clumsy—but remember, this is a first-cut design. Considerable refinement should occur before we implement it. Also, interfaces to the "outside" world which had been indicated in the dataflow diagrams by means of sources and sinks need to be addressed. This is accomplished by means of interface modules which are created solely for this purpose. Datastores are indicated as files which are accessed by the modules which create, modify, and/or use the information in them. Other interface modules that are needed are modules to PRINT or produce reports which were not included in our original analysis.

The basic process by which the Structure Chart is derived from the dataflow diagram is presented below:

Identify the highest level for conversion.   As discussed in Chapter 3, many levels of detail are possible in dataflow diagrams. The most common one to use in defining a design is level '0'. It is the highest level at which meaningful detail is present. If level '0' is not very useful or is overly simple, then replace the most complex processes with their surrogates. Be careful not to create a new level '0' which is overly complex. Also, how will we know if the level '0' we have is overly simple? We won't for sure until we assume it isn't and go through the remainder of this procedure and obtain poor results.

Identify an appropriate "boss" module.   Remember that, during this initial conversion step, each process bubble will be translated into one module. It may not still be a single module when we are done with the process, but this is just a first cut at a design. The problem we are faced with in constructing even a first-cut design is just where does the "boss" module come from? During the Structured Analysis we created a set of processes formed into a hierarchy such that one (the context diagram) could be decomposed into others, and those into others, and so on until functionally primitive processes were encountered. In such a hierarchically ordered network no process really qualified as a "boss" in that it controlled the others. Faced with this situation, we really have only two choices: install a "boss" which was not part of the original network, or elevate one of the processes-turned-modules to the position of system boss.

Draw a preliminary Structure Chart.   Once the boss module and its immediate subordinate modules have been identified via the preceding two steps, draw the lines of control into a Structure Chart using module symbols in place of process symbols. The name of the modules follow the same conventions that the naming of processes did. Hence, we will initially use the same names (transitive verb, object noun). Later, or as a result of some project standard or naming conventions, we may wish to change this name. In addition to the processes at the context and '0' levels, we need to incorporate the processes at other levels into our chart. This is done by taking each process/module at level '0' which has surrogates and treating them as subordinates on the structure chart. If any of them have surrogates, we do the same thing. We continue with this until all the processes defined during the analysis have been included (Figure 10.6-2). The overall Structure Chart would be the combination of the system level and all surrogates (Figure 10.6-3).

Complete the Structure Chart.   At this point we have created a Structure Chart which shows the control lines (calls) but does not include the transfer of data via data couples or the transfer of control information via control couples. We include both at this point. First we address the matter of the data couples. We note that the surrogates of an archetype process exchange data among themselves. Since they are now controlled by a boss module and since communication of data will

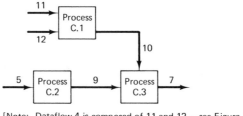

[Note: Dataflow 4 is composed of 11 and 12 — see Figure 10.6-1]

Level-'1' DFD for Subsystem C

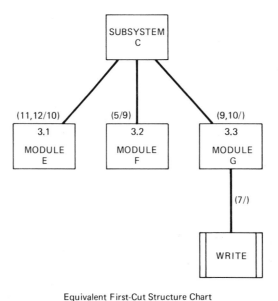

Equivalent First-Cut Structure Chart

Figure 10.6-2:   Inclusion of Remaining Processes

usually take place via a CALL, we have each module send the data destined for a fellow surrogate to its boss, which in turn passes it on. This may seem awkward—but remember, this is an early stage, and a lot of refinement is to come. Our primary concern at this point is that we do not lose anything we may have had in the analysis. The next issue we address is that of the control couples. These are inserted as we inspect modules and their signaling needs and relationships. Remember, any data or control couples that we create to address some discovered need or error must be defined in the dictionary.

At this point we have a design (of sorts) which is only partially complete, unrefined, and basically frightful! This model of our system is just a starting point. It is *not* a design which should be implemented without first going through a considerable amount of refinement. The nature of the refinements and their evaluation is the subject of later chapters.

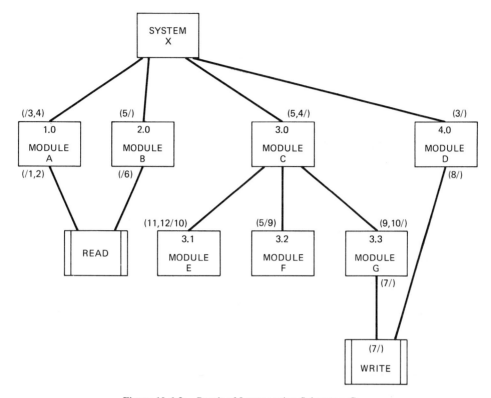

Figure 10.6-3:   Result of Incorporating Subsystem C

10.6.1 Select-a-Leader Approach

This approach involves determining which of the various bubbles on a dataflow diagram would be the most appropriate choice as a leader or ''boss'' module. The technique is to cut or break off the dataflows which are clearly input oriented and those which are clearly output oriented (Figure 10.6.1-1). What is left,

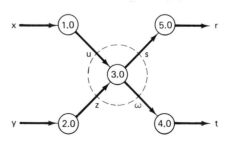

Figure 10.6.1-1:   Example of a Dataflow Diagram with Input and Output Flows

Leader candidate — neither input nor output oriented — in the "middle"

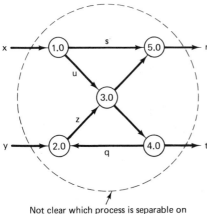

Not clear which process is separable on
the basis of input/output orientation

Figure 10.6.1-2: Example of a More
Common Type of Dataflow Diagram

supposedly, is *the* module best suited to be the leader. It is the *center*. Unfortunately, it usually turns out that one cannot always locate the center or transform/
transaction center for the design (Figure 10.6.1-2). When this happens, we "hire"
a boss. That is, we create a new module which does not have a pedigree in the
original analysis and let it be the boss.

10.7. SUPPLEMENTING STRUCTURED DESIGN

During the Structured Analysis phase, we were advised to reduce or eliminate all
processes that related to getting or putting data, input, output, and printing. The
reason was that we were trying to develop a specification, *not* develop the system
prematurely. Now that we have developed a satisfactory analysis and performed the
translation process, several modules are not present. Specifically, reporting modules are nearly absent. As we proceed to identify and add these to the Structure
Chart, we discover that there is no pseudocode defined for them. We can either
make the pseudocode up as we go along (a common practice) or we can employ
some method which will aid us in defining this pseudocode. One method [2, 3]
which is particularly well suited for use in conjunction with Structured Design is
nearly ideal for defining this pseudocode.

The basis for this method is that the hierarchical structure of the data should
be used as a basis for the structure of the code (and its design). What we are
suggesting is to use Structured Analysis and Structured Design in its nearly original
form to obtain a first-cut Structure Chart, then complete the first-cut chart by
employing an additional technique. The procedure is as follows:

1. Identify all input data, noting any hierarchical relationships that may be
present.

2. Use the hierarchical relationships to form a nested set of diagrams depicting the hierarchy of data items of interest.

3. Document the relative rate of occurrence of each data item of interest with respect to the other data items of interest.

4. Assign each level documented in step 3 above a descriptive label.

5. Repeat steps 1 through 4, inclusive, for the desired output data.

6. Describe the program primitives necessary to accomplish the processing of input into output. To do this, first identify the read statements, the branch statements, computations, outputs, and subroutine calls.

7. Create a flowchart made up of symbols such as:

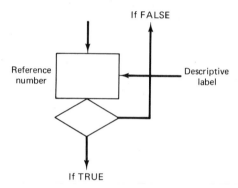

This "flowchart" will describe the sequence of the logic described earlier. Each process will have a BEGIN_PROCESS and an END_PROCESS block. Nesting will be accomplished via the use of branch statements.

8. Number each flowchart element and expand it to simplify wherever possible. Use only the basic instruction set (i.e., sequence and branch).

A more detailed explanation of the use of this method is available in [2, 3].

10.8 DEVELOPING A STRUCTURED DESIGN "WITHOUT BENEFIT OF A STRUCTURED ANALYSIS"

Although the ideal scenario is that a Structured Analysis is developed before a Structured Design is begun, it often happens that a Structured Design is attempted based on some set of "requirements" or list of functions. There are a lot of ways in which this can occur. The point is, it *can* occur. What should be done about it?

Lest it be thought that I am suggesting Structured Design be "Slam Dunked," let me observe, first, that what will be described here should be considered an emergency procedure, *not* standard operating procedure. It is a lot like CPR (Cardio-Pulmonary-Resuscitation)—a procedure to be used because the situation calls for it. Second, the results of this activity need not be of poor quality. They should put the design somewhere in the realm of reasonableness. The shortcomings of this "emergency procedure" will most likely be in level of detail and depth of understanding, *not* in consistency, completeness, and correctness. So much for preliminaries.

Perhaps the best way to describe this approach is to cite a specific instance of its use and draw some general rules from it.

THE INTEGRATED OFFICE PROJECT

A few years ago, a software project found itself in dire straits. The project was directed at the development of an integrated software package to support all aspects of general office work. The target environment was the IBM PC/XT™ or AT™ (both are trademarks of the IBM Corporation) hardware. One of many points of disagreement among project members was what sort of operating system would be appropriate. But more serious problems existed. Many of the project team members had previously been involved in development of one type of system or another that would be involved in this project. For example, word processor developers, spreadsheet developers, database management system developers, and people who had previously developed integrated systems were all involved on the project team. Each brought their own experiences, biases, and "vision" of what an automated system ought to be like. Naturally, there was little agreement on the details of this system. However, there was agreement on the high-level partitioning of the system. To some, this may seem odd, so an explanation is in order.

Earlier in this book we pointed out the problems that can occur in database development when we focus our attention on the *product* rather than the *information* involved in the system. We pointed out that the result of such an approach would be the creation of individual, independent databases which would each support some function rather than an integrated, nonredundant database capable of supporting *all* functions. A similar situation can exist with respect to the application code. It definitely existed in this case. The high-level or level-'0' view was such that the system had to perform some functions which did not seem to "belong" to any *single* level-'0' functional area. The level-'0' view held that this system would be composed of:

> Office automation functions
> Telephone functions
> Data processing functions

Each of these included other subfunctions. But some subfunctions (e.g., database management) seemed to "spread" across two or more functional areas. Others (e.g., terminal emulation and data sharing) did not seem to "belong" to any level-'0' parent.

This situation (and others) caused some dissention and disagreement within the team. The telephone functions included the ability for the system to log telephone calls, place them, administer telecommunications, and communicate with other systems, as well as other features. Office automation had been allocated the usual word processing, database, and spreadsheet functions as well as maintenance of the telephone directory. This last feature caused some additional discussion with the telephone group, since they felt this was something they should do. The data processing function would provide users with the ability to create applications either from "scratch" (i.e., native code) or by using an applications generation capability built into one or more of the office automation packages.

As you can see, there was considerable opportunity for the project to end up in a kind of "gridlock" wherein agreement did not occur and no progress could be made. Into this fray came a consultant preaching the need for analysis to precede design. At that point a "design" agreement was reached that *if* an analysis could be generated quickly and *if* it would result in an improved design, *then* do it.

Since we did not have the classic analysis situation in which interviews are conducted and the data reduced to a Lifecycle Dictionary, Dataflow Diagrams, etc., an alternative approach was used. It was obvious that the "top-down" approach, which many hold to be nearly sacrosanct, not only was inappropriate but simply would not work. What was used was a type of "cluster analysis." The team was asked to identify each and every task or function that the system would have to perform. Participants were admonished to disregard any preconceived notions they had about what the organization of these ought to be. All that we were after was a complete list. The following list was developed:

Configure telephone

Word processing

Maintain phone list

Database management

Electronic mail

Spreadsheet

Graphics (for financial reporting)

Emulate a terminal

Maintain calendar

Call accounting

The list contained some inconsistencies in naming, but it was a breakthrough. It represented a point of agreement and a beginning of our search for a new partitioning.

Do not get the impression that obtaining this list was easy. It was arrived at by reducing a much longer list. The reductions were accomplished through identifying and eliminating functional duplicates (even though the names were different), removing what were, essentially, mechanisms for accomplishing a task, and collecting low-level, detailed tasks together, replacing that collection with one appropriate name.

The next step was more difficult. It involved trying to identify the "natural aggregates" of functions within this list. In examining the level-'0' partitioning, the office automation area was both the "fuzziest" and potentially the most complicated. It was decided that this one was the key and would be addressed in the most detail. This is consistent with the experience we had in the Structured Analysis phase. In that phase, the most effective strategy was to identify the most complex process and decompose it, then to proceed to the next most complex and so forth. This is also a highly effective strategy throughout the development of a Structured Design. What we were looking for were office automation tasks that seemed to be related either by their use of the same data or by some functional characteristic. At first, we did not concern ourselves with what name should be applied to a cluster, just whether or not it seemed to be a reasonable cluster. The names for some clusters were obvious and instantly agreed upon. Others took some experimentation and negotiation. The clusters (and their components) that we ended up with were:

Automate Office Functions

Word processing
Database management systems
Spreadsheet
Graphics functions

Administer Telephone System

Administer telephone configuration
Account for calls

Transfer Data Between Users

Electronic mail
Terminal emulation

Administer Personal Planner

Maintain phone list
Maintain calendar

The above four functions or processes were used to replace the office automation process. This gave us a level '0' which contained the following processes:

Automate office functions
Administer personal planner
Process telecommunications
Produce financial reports
Generate programs

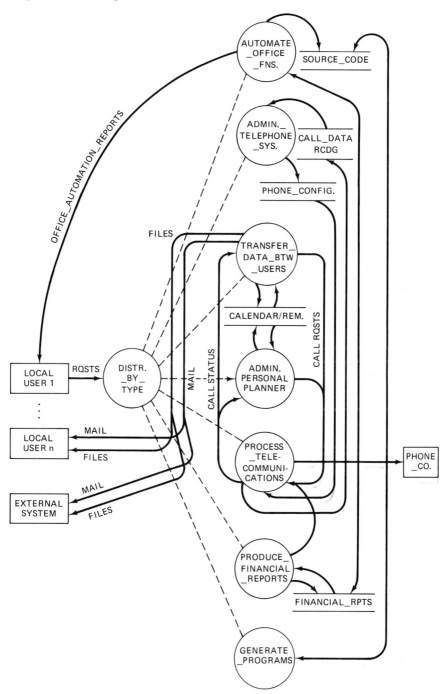

Figure 10.8-1: Level-'0' DFD for the Automated Office

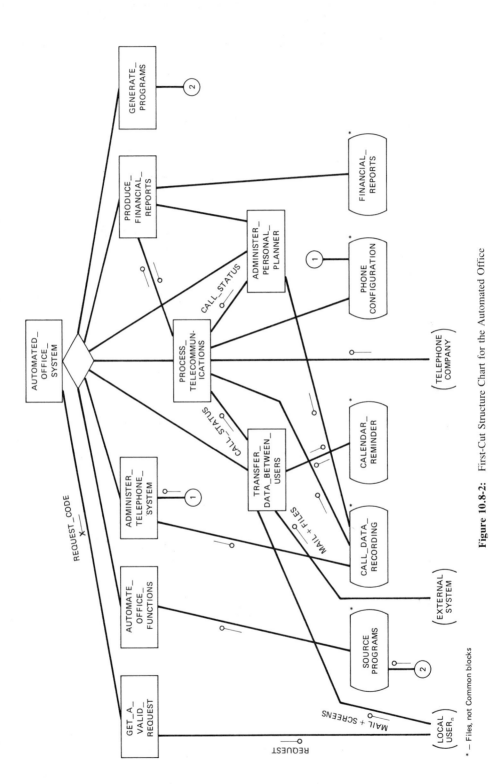

Figure 10.8.2: First-Cut Structure Chart for the Automated Office

* – Files, not Common blocks

The above were used to generate a level-'0' dataflow diagram (Figure 10.8-1). Note that since we have a real-time system, we have chosen to incorporate a dummy process called DISTRIBUTE_BY_TYPE to clarify matters. The DFD lead to the identification of dataflows, which were documented in a dictionary to the element level.

Inspection of the DFD in Figure 10.8-1 reveals that the naming is not all that it should be. For example, what does "ADMINISTER" mean? It is not clear how this and some of the other processes actually transform information. The naming here indicates a certain amount of haste and a lack of refinement but that is exactly the situation we are in. However, this DFD has given us some insight into the partitioning of the problem. It will also help us obtain a first-cut Structure Chart from which we can begin refinement. We are working in haste in this situation but we are trying to get the *best* structured design we can under these constraints. An examination of the Structure Chart reveals a lack of detail. The refinements in this mode of operation will occur here—in the design phase. In such situations each module should be factored into subordinate tasks/modules. Also, many of the data couples are unnamed as a result of rushing out of analysis. This rapid design scenario causes us to run the risk of poor module quality unless multifunction modules are broken into single function ones.

The developed DFD and subordinate DFDs were used to derive a Structure Chart. It differed significantly from the one derived using the earlier level '0'. The differences were mostly in the areas of "tramp data," decision splits, and levels of cohesion (see Chapters 12 and 14). The dictionary was further expanded to incorporate design information.

The general process and principles that we can extract from this scenario are:

1. Imposing a "functional partitioning" on system decomposition does not lead to a well-partitioned DFD or Structure Chart.

2. An "emergency" approach to obtaining an adequate partitioning is to list, in detail, everything the system is to do, then look for "natural aggregates." Each item would be stated as a verb/noun combination.

3. Reasonable partitioning results in a set of processes which:

Provide complete coverage (i.e., no task, however small, does not logically belong to one of the resulting processes).
Are *true* transformers of data (i.e., we can actually follow dataflows into, through, and out of the system).

The resulting *first-cut* Structure Chart derived from Figure 10.8-1 is presented in Figure 10.8-2.

10.9 ANALYSIS TO DESIGN—AN IRREVERSIBLE PROCESS

A question that often comes up is whether or not the dataflow diagrams that were developed during analysis should be modified as a result of "discoveries" made during the design phase. The key factor we need to take note of to resolve this question is the *basis* for each of these results (i.e., dataflow diagrams and structure charts).

The dataflow diagram was the result of a process of cataloging inputs and outputs and identifying transformations processed through a tracing and backtracking activity. The goal of the activity was a model of *what* the system must do. The data descriptions, event model, and information model all embellished this result but it is still the same—a specification.

The goal of the design phase is the creation of a system blueprint. Even in the logical design activity, the basis for this blueprint is hierarchic and control-oriented. In physical design, we add operational constraints to further refine our results.

The transformation of the analysis results into a design is *not* reversible for several reasons. The foremost of these is the fact that an important shift in our frame of reference occurs when the dataflow diagram (which is a non-hierarchic leveled network) is transformed into a structure chart (which is hierarchical, non-network-oriented, control-based). This shift in perception is like a non-linear transformation of coordinates. There is no way back. Besides, if we went back to the dataflow diagram and added/deleted processes and dataflows to match the structure chart, all we would have is another structure chart in a different format.

The advice that software developers and maintainers should adhere to is that the blueprint resulting from physical design should be kept up to date but the dataflow diagrams should not match it one for one.

10.10 THE TRANSITION FROM STRUCTURED ANALYSIS— SUMMARY

The problem with making such a fuss over types of systems—arguing about how to recognize one type or the other, and about ''afferent'' and ''efferent'' flows (that's right folks, these terms were introduced in at least one text—they are terms from biology which I am sure you will all recognize)—is that we might lose sight of where we are attempting to go. The purpose of the translation process is to get to a starting point from which we can refine and revise until we get what we are after. What we are after is a quality design. Remember, design is an amoral activity. As such, any means to the desired end appears to be OK. Spending too much time and energy on deciding the subtle differences between one type of flow and another can cause us a great deal of pain without much to show for it. We are looking to develop an architecture from which a quality system will emerge.

REFERENCES

1. T. De Marco, *Structured Analysis and System Specification*, New York: Yourdon Press, 1978.

2. J. D. Warnier, *Logical Construction of Programs,* 3d ed., trans. B. Flanagan. New York: Van Nostrand-Reinhold, 1976.

3. K. T. Orr, *Structured Systems Development.* New York: Yourdon Press, 1977.

CHAPTER 11

Structured Database Design and the Lifecycle Dictionary for Design

In this chapter we do not attempt to address the subject of database design at a low level of detail. Rather, we introduce it in sufficient detail to direct the reader to further investigate specific topics, as necessary. The presentation here completes the spectrum of topics treated in this book from analysis through design.

The Lifecycle Dictionary entries that were developed during the analysis phase concentrated mainly on the generic composition of information rather than its physical characteristics. For example, in the case of an interrupt or event which caused the system to change its mode of operation, we tended to direct our attention to describing the compatibility of this event with others rather than on defining its bit pattern. Similarly, we ignored the physical format of information such as a customer's name, address, and telephone number, since it was not of primary interest. At this point, we have developed a Process Model. Does this mean that we must immediately turn our attention to the definition of the dictionary supporting it in terms of physical characteristics? Definitely not!

Remember, Structured Design will go through two major phases. The first is the Logical Design phase. It involves the description of the design in abstract terms. That is, it is directed at describing the system from a design standpoint but without regard for the details of the operating system it may have to run under, the programming language it will be implemented in, the timing and sizing constraints, or other implementation factors. This phase is immediately followed by the Physical Design phase, during which we resolve all of those practical, physical issues. Similarly, the

Lifecycle Dictionary during the Structured Design phase undergoes a two-step development.

We will discuss the phases of the Lifecycle Dictionary during design as well as the differences between the logical and physical aspects of the Lifecycle Dictionary during this phase.

11.1 OVERVIEW OF STRUCTURED DATABASE DESIGN

Database design has some parallels to the more classic code or software design. It goes through two stages: Logical and Physical. For our purposes we will define database design as a process directed at transforming subject database specifications and E-R-A models into database schemas which can be implemented. This facet of Structured Design has been ignored by most [1, 2] authors.

What we are seeking in the area of Structured Database Design is a description of the database which is needed to support the application description. How this database gets implemented is not really our concern. We are attempting to state what the logical design of that database is. Our letting the Data Administration (or related group) perform the implementation, integration into the corporate database, and/or refinements ensures that we will not end up with a series of individual, product-oriented or application-oriented databases which are not integrated. The use of a series of databases is popular with most project managers, programmers, and vendors of mass storage devices. One can see that this situation arises easily. As each application system is built, a database is built to support it. Over time, several of these exist, often containing redundant information.

So, what is so "bad" about having redundant data? Data storage is getting cheaper all the time. Why worry about it.

One major problem with redundant data is that this information is not under control—as illustrated by the following experience:

> A manufacturer of computing equipment experienced phenomenal growth in its early years of operation. As a result, they had to develop a database to support the sales organization's efforts. This was quickly followed by the development of an application and database to support the accounting function. As time went on, separate databases were developed for the shipping/receiving department, the manufacturing department, and the service department. The situation that developed caused problems so severe that a consultant was brought in. The data in the various databases was inconsistent and was under no type of charge control; individual departments were run like fiefdoms. A customer would purchase or (more likely) lease a piece of equipment. Manufacturing would be notified manually, accounting notified manually, and shipping/receiving notified manually. The equipment would be manufactured (most items previously manufactured were already leased) and shipped. Often the bill would go to one address, the equipment to another—and often one or both of these addresses would be wrong. This was particularly true of clients with multiple facilities and

those that had moved since the last transaction. The consultant offered a recommendation to form a corporate data model and eventually develop a database from which individual needs could be served. It fell on deaf ears. The consultant was terminated. Unfortunately, the problems with billing got so bad that cash flow problems "ate" the company and it went into Chapter 11 a year later.

Most of us are not dealing with information problems that affect an organization as directly and dramatically as this one did. However, the cumulative effect of local problems can be nearly as serious.

11.2 APPROACHES TO STRUCTURED DATABASE DESIGN

As in the area of code design, a number of different techniques can be employed. These can be summarized into three classes of approaches:

> **Top down.** This involves a two-stage process in which the E-R-A model is refined, then transformed into a relational schema.
>
> **Bottom up.** This involves the specification of processes and user views followed by integration of the user views.
>
> **Middle out.** Like the top-down approach, this begins with refinement of the E-R-A model. This is followed by integration of design view and the generation of relational schema.

11.3 DATABASE DESIGN AND SA/SD

The Structured Analysis package that was developed as part of the activities described in Part II of this book contained several items of interest to Structured Database Design, including

> Datastores, sources, sinks, data item definitions, and E-R-A models with descriptions (i.e., definitions of relationships between entities and policies about entities)
>
> Level-'0' descriptions of all data flowing into and out of the system.

We will use this material to develop our Structured Database Design in the following way:

> Use the datastores to identify the subject databases.
>
> Use the E-R-A model to define the allowable dataflows in/out of datastores.
>
> Complete/verify the Data Access Model (dataflows in/out of the subject database datastores map into data access specifications)—see Section 11.4.
>
> Use the refined data access model to provide the abstract database access modules needed as part of the Structured Design.

11.4 DATA ACCESS MODELING

In general, every database is subject to accessing by several application programs. A similar situation exists on a smaller scale with respect to an assemblage of modules and the database (or portion of one) that supports them. These accesses, and others, reflect the enterprise or business activity the system is to support. The data most important to the enterprise is likely to have been incorporated into the E-R-A model. A Data Access Model (or DAM) is a process-oriented view of the database being developed [S3]. It is, essentially, the link between the database and the application modules.

The DAM is developed through the use of portions of the Process Model and the Information Model. Specifically, the dataflow diagram and the E-R-A diagrams are used. The process for defining a DAM is:

Determine entities of interest to process.

Isolate the corresponding portion of the E-R-A model.

Remove any remaining unneeded data elements.

Create a DAM by documenting the sequence of accesses.

Describe the access logic using pseudocode or structured English.

Document specifics about the access, as necessary, to aid in physical design.

11.5 EXAMPLE OF DAM DERIVATION

As an example of how to develop a DAM, let us reexamine a variation in our Student-Teacher E-R-A diagram (Figure 11.5-1).

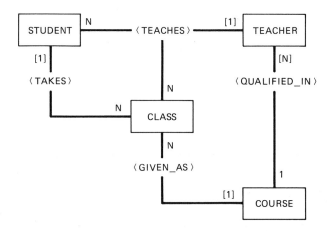

Figure 11.5-1: Partial E-R-A Model to Demonstrate DAM Development

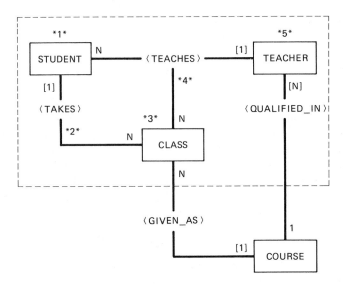

Notes: *n* indicates access sequence number. Access sequence is 1-2-3-4-5.

Figure 11.5-2:   Required Accesses

Let us assume that our system is going to be required to generate a report on the students who have the same teacher for more than one class. This would mean that a series of accesses would be required and in a specific order (Figure 11.5-2). To complete our description of the DAM we need to identify [3] the information listed in Table 11.5-1.

TABLE 11.5-1:   Information Necessary to Develop a DAM

Access Fields
What fields are used to access the data

Sequence
The order, if any, in which data are accessed

Frequency
The absolute or relative rate at which a given set of data is accessed

Numbers
Values and ranges for variables

Iterations
Any cyclic accessing

Edits
Validation of data

Much of the above information is incorporated into the Lifecycle Dictionary.

11.6 UTILITY OF DAMs

The transition from this logical view (composed of the E-R-A model, the DAM, and accompanying Lifecycle Dictionary entries) to a physical database design is relatively simple when the database management system (DBMS) that will be employed is a relational one. If the DBMS is a network or hierarchical type, the transition will be more difficult. It will require revision of the DAM during physical design. What we are attempting to do is to implement the database in such a way as to make it application independent.

An appropriate analogy is that of the specialty food shop vs. a supermarket. Without the incorporation of an overall database design, application development will result in the creation of individual databases, each supporting a *simple* application (or, possibly, a small number of applications). This is analogous to creating a specialty food center for each type of taste in the community. One might end up with several shops, such as a Kosher delicatessen, an oriental food store, a bakery, a Mexican food delicatessen, an Italian food store, and so on. (If your favorite is not in the list—sorry.) Hence, to obtain the materials to make meals for a week one would have to go to the butcher shop, the bakery, the coffee and tea shop, and other food specialty stores. An alternative approach would be to include all of these types of foods into some integrated marketplace—a supermarket. Although the supermarket would not meet *any* needs ideally, it would meet *all* needs satisfactorily. What we should be striving for in information modeling and database design is the creation and maintenance of an official information supermarket which meets the needs of the enterprise.

11.7 THE DESIGN LIFECYCLE DICTIONARY

The main changes to the Lifecycle Dictionary during the design phase are the incorporation of flags, the addition of any data elements inadvertently left out during the analysis phase, the correction of errors, the addition of ''mailboxes'' (see Chapter 9) for signaling between/among modules, and the incorporation of pseudo-code and data items for modules that were created as a result of design (e.g., utilities, data interface modules, input, and output modules).

Just in case there is any doubt, *flags* are dictionary items which are not inherent in the enterprise. They owe their existence to the fact that the solution to the problem was *implemented*. Data items exist owing to the nature of the enterprise. Hence, flags are inherent in the solution, and data items are inherent in the problem. Examples of flags are presented in Table 11.7-1.

TABLE 11.7-1: Examples of Some Common Flags

Typical Flag Name	Comment
ITEM_VALID	Used with edits
OUT_OF_LIMITS	Signals a condition
END_OF_FILE	Signals an event
MISSILE_RDY	Signals a condition

The information that must be added to the Lifecycle Dictionary may take any of several forms. The *minimum* information required is shown in Table 11.7-2.

The list in Table 11.7-2 may be expanded or contracted depending on the nature of the project, the standards and guidelines employed by the development team, and personal preference. Concern for those who will have to maintain this system should guide the development team's thinking. What information will they need?

TABLE 11.7-2:   Description of Items Used in the Design Dictionary

Item	Descriptive Information Required
Module	Name Author Number Date created Date of last revision Purpose Pseudocode Called by Calls Modules synchronized with Data used/how obtained Flags used/how obtained Type of module
Mailbox	Name Author Number Date created Date of last revision Purpose Module(s) suspending to Module(s) resuming to Key Algorithm/conditions for suspension/resumption
Syncbox	Name Author Number Date created Date of last revision Purpose Module(s) suspending to Module(s) resuming to Algorithm/conditions for suspension/resumption
Data couple	Name Author Number Date created Date of last revision Used by Created by or part of
Control couple	Name Author Number Date created Date of last revision Used by Reset/changed by Purpose

11.8 PSEUDOCODE

In Part II of this text, we discussed pseudocode from the standpoint of its role during analysis. In this section, we will discuss it in terms of its role during the design phase. Not all of the various constructs described herein will be available or possible in all language and machine environments. However, many of these are supported directly or their equivalent can be created indirectly.

Pseudocode is a means of describing policy that we plan to incorporate into the final code as a cost-effective means of creating *accurate* documentation of what a module does. At the present time, there are no published standards for use with pseudocode. The need for a standard for pseudocode arises from our desire to standardize. Productivity is greatly reduced when one has to adjust to or accommodate one software engineer's style of pseudocode after employing their own or examining someone else's style of pseudocode. The purpose of this documentation is to detail a set of pseudocode standards which meet three needs:

> To describe the algorithms that will be implemented in a way that will facilitate the process of validating and implementing the algorithm.
>
> To document, in an understandable way, a procedure. This is particularly useful for assembly language routines.
>
> Creation of a standard which will enable the Lifecycle Dictionary tool to more easily and accurately store *and analyze* pseudocode and correlate it with other design elements.

There are two classes of standards for pseudocode—conceptual and notational.

11.8.1 Conceptual Standards for Pseudocode

- Pseudocode describes policy but is intended to simplify and clarify policy.
- Policy may take any of several forms, including mathematical equations, textual statements, and codelike procedural descriptions.
- During the physical design and implementation phases, pseudocode will incorporate how the processes are carried out.
- It must be capable of specifying, or at least implying, all the capabilities and features of the physical design.
- It is recommended that pseudocode be developed for each process in the dataflow diagrams that are part of the analysis. As a minimum, pseudocode should be created for the primitive-level processes in the analysis model.
- Pseudocode shall be developed/revised for each module.
- In all instances, the element(s) that will be used should be referred to rather than the primary data item of which it is a part. The intent here is to reduce or

eliminate those instances where a module receives an entire record, for example, but uses only some subset of that information. This will result in "tramp" data situations which will lead to higher maintenance costs and greater difficulty in adapting the system to new requirements. If this happens frequently, it usually means that the choice of what should or should not be contained in such a record was not prudent. That is, a subset of those elements actually share a relationship which has not been identified. It is suggested that in such cases, the record (or other structure) involved be broken into two or more substructures and the feasibility of this in implementation be investigated.

- *No* "magic" numbers are to be embedded in the pseudocode. This means we will not use literal values in the pseudocode but rather will completely parameterize it using meaningful variable names. The only exceptions occur in indexing situations (e.g., Do ~~~~ $i = 1$ to $i$max). In such cases, the starting value of the index (e.g., 1) does not have to be defined in the dictionary *if it has no special significance*. However, the limits of the loop (e.g., $i$max) should *not* be a literal but be a variable which is defined in the dictionary.

- The pseudocode for a given module in design must include a CALL statement for each and every module that is called by the module being described. All logic required for the module to properly perform its tasks must be included. The CALL statement(s) must include enumeration of the data that is sent from the calling module to the called module and the data that is returned from the called module to the calling module.

11.8.2 Notational Standards for Pseudocode

The names of the data items used in the analysis and the names of the data couples and records used in design are the *only* link tying the analysis and design together. For this reason, the variables or data items which are used in the pseudocode must be defined in the Lifecycle Dictionary. The data element names incorporated into the pseudocode must be the *same* names used during the analysis phase and defined in the Lifecycle Dictionary. The only exceptions are cases where what was a single record in the analysis becomes two or more records in the design, control couple names (which, for the most part, exist only in the design), new data items introduced to accommodate some oversight or requirements change, and name changes necessitated by the maximum size allowable for variable names with this compiler (in this case, 15).

The pseudocode must be indented n columns to indicate nesting levels or control gates necessary to reach a particular section of it, where n is equal to or greater than two.

The pseudocode for a given process (or module during the design phase) should not exceed approximately 100 lines (i.e., two typed pages), single-column format.

All operands (e.g., WHILE DO, CALL, CASE) must be entered as all capitals.

11.8.2.1 Constructs

The constructs to be used within the pseudocode are:

```
                    IF THEN ELSE

                    WHILE DO

                    UNTIL DO

                    SELECT CASE

                    SEQUENCE

                    BLOCK

                    ASSIGNMENT

                    FOR LOOP
```

Each is discussed in more detail below. It should be noted that many other formats are possible for these constructs.

11.8.2.2 The IF-THEN-ELSE construct

If IF test is to be entered as follows:

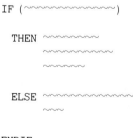

where the condition is contained within parentheses.

The content and format of the IF test condition may take many forms. The set of operands to be used in order to gain the desired expression is presented below:

Operand	Symbol
Equal	EQ
Less than	LT
Less than or equal	LE
Greater than	GT
Greater than or equal	GE
Not equal	NE
Or	OR
And	AND
Set membership	IN

11.8.2.3 The WHILE DO construct

The WHILE DO construct is to be entered as follows:

```
WHILE (^^^^^^^^^^^^) DO

        ~~~~~~~~~~
        ~~~~~~~~~~~~~~~
        ~~~~~


ENDDO
```

where the condition is contained within parentheses.

11.8.2.4 The UNTIL DO

The UNTIL DO construct is to be entered as follows:

```
REPEAT

       ~~~~~~~~~
       ~~~~~~~~~~~~~~~
       ~~~~~


UNTIL (^^^^^^^^^^)
```

where the condition is contained within parentheses.

11.8.2.5 The CASE construct

The CASE statement is to be entered as follows:

```
SELECT USING (variable_name)

CASE (~~~~~~~~~~~~ OP ~~~~~~~)
      ~~~~~~~~~~
      ~~~~~~~
```

```
CASE ( ~~~~~~~~ OP ~~~~~~ )
        ~~~~~~~~~~~~~~~~~~
        ~~~~~~~~~~~~~~~~~~~~
        ~~~~~~~~~~~~~~~~~~~~~~~
        ~~~~~~~~~~~~~~

CASE ( ~~~~~~~ OP ~~ )
        ~~~~~~~~~~
        ~~~~

OTHERWISE
        ~~~~~~~~~~~~~~~~~~~~
        ~~~~~~~~~~~~~~~~~~~~


ENDSELECT
```

where

variable_name	is the name of the variable being used to make a selection from among the various cases.
OP	is the name of the operation being tested for. For example, in TARGET_ID EQ UNKNOWN, EQ would be the operation.
OTHERWISE	is used to denote the situation in which *none* of the CASE conditions are met. The instructions which follow the OTHERWISE indicate what the module will do when no other case is met.

11.8.2.6 The SEQUENCE construct

This construct is self-explanatory. It requires no indentation in and of itself but may be indented as a block, owing to the presence of one or more conditional statements preceding it.

11.8.2.7 The BLOCK construct

The use of this construct is demonstrated below:

```
BEGIN BLOCK

        ~~~~~~~~~~~~
        ~~~~~~~~~~~~~~
        ~~~~~~~~~

        IF ( ^^^^^ )
          EXIT BLOCK
        ENDIF
        ~~~~~~~~~~~~~~

        ~~~~~~~~


END BLOCK
```

where the EXIT BLOCK causes execution to transfer immediately to END BLOCK.

11.8.2.8 The ASSIGNMENT statement

An example of the use of the ASSIGNMENT statement is:

```
DESTINATION := SOURCE
```

11.8.2.9 FOR LOOP statement

An example of the use of the FOR LOOP is presented below:

```
FOR (VARIABLE) := INITIAL_VALUE TO END_VALUE DO

    ~~~~~~~~~~~~~~~~
    ~~~~~~~~~~~~~~~~
    ~~~~~~~~~~~~~


ENDFOR
```

11.8.3 The CALL Statement

The CALL statement is used to indicate that one module is relinquishing control to another module. This CALL statement is to be entered as follows:

```
CALL module_name (ip1,ip2, . . , ipn/op1,op2, . . . ,opm)
```

where

module_name	is the name of the module being called
ip$i$	are the parameters that are being sent to the module being called
op$j$	are the parameters that are being returned from the called module

11.8.4 Other Commands

Often there is a need in real-time systems to set up synchronizing arrangements between/among modules. That is, often in multiple-processor situations one module must "know" whether or not another module has successfully completed a task. The typical way to do this is to use CALLs and control flags. However, this proves too expensive from an execution-time standpoint. Other ways to accomplish this and to make use of special features of the compiler intended for real-time applications are described below.

11.8.4.1 Mailboxing

A *task* is defined to be a procedure which can be executed in parallel. It has no parameters. Input/output is accomplished via mailboxes. Mailboxes are *not* used for efficiency. They *are* used for the purpose of decoupling of control between tasks— that is, to provide communication between tasks which does not have critical timing dependencies. A mailbox passes a message, not a flag. The message has content, even though it may be a single word. This word can be passed as a pointer, access a token, or serve other purposes. It can *logically* transfer any volume of data. It provides a means of:

> **Guaranteed contention resolution.** No matter how many tasks are waiting on a message, only *one* receives it.
>
> **Handshaking.** It replaces that same function that is taken for granted in a procedure call.
>
> **Timing independence.** The receiver may ask for data *before* it is sent and still get it (when it is sent).

There may be exceptions, but they must be considered on a case-by-case basis. The notation used to describe mailbox situations is described below.

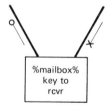

where to use this arrangement we may employ the operators

```
PEND (MAILBOX_NAME, TIMEOUT/MESSAGE,ERROR)

POST (MAILBOX_NAME, MESSAGE/ERROR)
```

and

> mailbox_name is the identifier for the mailbox used
> error returns 'no error', 'timeout', mailbox full, or mailbox empty

11.8.4.3 Reentrant modules

Reentrant modules are to be identified as such by means of the following notation:

```
PROCEDURE NAME
REENTRANT
COMMENTS (explaining reasons for this)
```

11.8.4.4 Compiler options

Several compiler options are available. An example is shown below. In every case, the reason for using such specialized features should be explained via comments.

```
COMPILER OPTION 'AMNESIA'
Comments explaining why used
```

11.8.4.5 Task SUSPEND–RESUME

Sometimes we need to have two or more modules signal each other using a technique which does not involve a COMMON block, since no data is passed between them. That is, the status of each module and its relationship to the other is handled through a mechanism which is transparent to the modules. Module A may have to SUSPEND itself at some point in its processing if Module B has not successfully completed some task or subset of a task. When Module B has completed, it issues a RESUME (or whatever the language construct is for the given machine), and Module A resumes its processing. If Module B issues the RESUME before Module A issues its SUSPEND command, Module A will continue without SUSPENDing. Often, it is required that if the suspended module does not resume within a specified time, then an error has occurred and remedial action must be taken. The use of SUSPEND and RESUME type commands is presented below:

```
SUSPEND (Module i/n/m)
```

where

Module i	is the name of the module being SUSPENDed
n	is the numerical identifier of the signal box as it appears on the Structure Chart
m	is the number of microseconds that can elapse before a malfunction shall be declared.

```
RESUME (Module j/n)
```

where

Module j	is the name of the module being RESUMEd
n	is the numerical identifier of the signal box as it appears on the Structure Chart

11.8.5 Incorporation of Test Cases

The creation and use of valid test cases begins with the identification of the minimal test set. It is recommended that this minimal set be composed of that set of data that is required in order to cause each and every branch point in each module in each outcome possible. For example, a simple IF test would require two test cases—one for a true outcome, the other for a false one. Modules which contain several sets of IF tests would require some combinatorial set of tests to accommodate tests which are precluded by or dependent upon other tests. It is recommended that the pseudocode contain a table of test range values as shown below:

BRANCH POINT	V1	V2	V_i
IF TEST #1	<0	–	>1
.			
.			
i			

where

V_i is the name of the variable involved (e.g., azimuth)

IF TEST #1 is an identifier for a given branch point

A chart such as the one above describes a test case

11.8.6 Pseudocode and Lifecycle Dictionary Summary

Pseudocode has long been received as a way to describe executable code in easy-to-understand terms. This has worked in many cases. If one accepts the premise that programming is an attempt at communication between alien species, then more than a description of executable code is needed. This additional information is specific to the project and organization but its role is the same—ensure cost-effective development and maintenance.

The Lifecycle Dictionary assists in accomplishing this by providing a central repository for all such information. Centralizing this information has the additional benefit that analysis, validation, and configuration control, and reuse on other projects are also possible and practical.

REFERENCES

1. E. Yourdon and L. L. Constantine, *Structured Design*. New York: Yourdon Press, 1968.

2. M. Page-Jones, *Structured Systems Development*. New York: Yourdon Press, 1978.

S3. Seminar, "Structured Data Base Design," developed by H. Walker, May 1986, offered by Software Consultants International, Ltd., Kent, Washington.

CHAPTER 12

Evaluating and Improving Module Interactions

Structured Design views software at two levels—the system level and the module level. Its goal is to arrive at a system architecture which combines high-quality, reusable modules in a simple way. That is, a quality software system is viewed as being composed of strong pieces which are linked in a flexible, simple manner for strength and durability. The first Tacoma Narrows Bridge failed not because the components or materials were faulty but because they were linked in a way that made that bridge susceptible to undamped harmonic oscillations. In a similar way, poorly designed software systems fail owing to their inflexibility in the face of change. No one builds them to fail. They just unwittingly create relationships that result in failure when the environment or input changes.

In this chapter we will discuss design evaluation at the system level. The smallest unit within the software system design that is discernible at this level is the module. This macroscopic or intermodule level of interaction is referred to in Structured Design as *coupling*. The coupling concept accepts the fact that there are many ways in which modules may interact. These are classified according to the degree of independence from each other that the modules exhibit. Within the Structured Design method, these interactions are partitioned into seven levels of coupling. This expanded list of coupling levels includes some real-time relationships. These are ranked in a way that supports the overall philosophy of Structured Design. This philosophy encourages simplicity and explicitness in system design. Hence, within this seven-level evaluation scheme, the simplest relationships between modules are deemed to be the most desirable while the more complex and obscure ones are the least desirable.

In this chapter we will examine each level of coupling through the use of examples and counterexamples. In most cases, we will find that our quest for improved coupling will result in unforeseeable improvements in the overall design as a side effect. For example, in general, simplifying a complex module relationship will often result in the creation of one or more simpler modules. Often these will be identical to others in the system, thus revealing a likely candidate for reuse as a utility module, or for incorporation as a system service, that might have otherwise gone unnoticed or been duplicated. The unforeseen benefit of this is the externalization of design decisions.

12.1 THE CONCEPT OF COUPLING

The term ''coupling'' refers to the nature of the relationship between two modules. From a design and maintenance standpoint, the simpler this relationship, the better or more desirable it is. Similarly, the more complex and intimate the relationship, the less desirable it is, since a strong interdependence bonds the two modules involved. When change is necessary, changes to modules which are strongly bonded often result in unexpected and undesirable side effects. For example, which of the following module relationships is less likely to cause development and maintenance difficulties?

1. Module A calls module B and sends it some data. Module B uses the data to compute a result. Module A receives the computed result (e.g., summaries, statistical analysis) in a returned parameter list.

OR

2. Module A branches to a location within module B, executes the next 48 words of instructions in module B, and transfers control back into itself.

Relationship 1 would seem to be the easiest to change and understand. Relationship 2 actually occurred in several places in a satellite command and control system. One programmer in particular engaged in this practice in order to meet the memory allocation requirements which had been placed upon him. This practice could and did cause undesirable ''ripple'' effects. When module B was changed, the programmer responsible for module B was unaware that module A used part of it in this way. Programmers who set up situations like relationship 2 rarely let others know about them before they leave the company. In this case, the launch date for the satellite system had to be ''slid'' while the problems were fixed in a ''fire drill'' mode. This created other problems. The main idea to keep in mind about coupling is that simpler is better.

Most experienced software engineers realize that simple interfaces make for simpler systems, easier to develop and maintain. Where disagreement exists is in just what (other than the two extremes cited above) constitutes a simple or desirable interface. The use of Structured Design greatly clarifies this problem by defining several levels of coupling, providing a simple mechanism for evaluating which type is present, and ranking them in order from most desirable to least desirable. Although some software designers may not agree with its order, this ordered list does provide an objective means of evaluating design decisions. Since design is a process of selecting from among alternatives, an evaluation scheme of this type is essential for consistent, objective design decision making.

The main value of the coupling concept in Structured Design does not lie in the fact that simple interfaces are viewed as desirable and complex ones as undesirable. This conclusion one could have arrived at a priori. No, the main benefit to us as software designers is that Structured Design goes considerably further in the refinement of this basic idea. This enables us to make objective evaluations regarding the quality of a design. Since design can be described as the process of selecting from among alternatives, there is a need for a value system against which to evaluate these alternatives. Whether we are dealing with an individual software design engineer or a team of them, the consistent and disciplined application of a common, effective value system is required for quality results [1]. In this way, the effects of the Structured Design method go beyond an individual program or system to the organization itself.

Given that some analysis method has provided an appropriate partitioning, software design really poses two kinds of problems to the software engineer: (1) to generate alternative software designs, and (2) to select the best one. Since design can become a never-ending activity [2, 3], we need to be able to detect when further refinements are not improving matters. The concept of coupling can aid us both in providing much-needed objective, quality criteria and in enabling us to detect when improvements are marginal or we have regressed.

12.2 LEVELS OF COUPLING

As we have pointed out, several different kinds of module interfaces can be employed in our software design. These range from simple to complex [4]. The levels of coupling are listed in Table 12.2-1 in order from simplest (most desirable) to most complex (least desirable).

TABLE 12.2-1:   Advisability of Various Coupling Levels

TYPE OF COUPLING	COMMENTS
Synchronization	Good—if appropriate
Informational	Good—if appropriate
Data	Good
Stamp	OK
Control	Be careful!
Common	Avoid this
Content	Outlaw!!

In the sections which follow, we will define each of these levels, demonstrate what it looks like on a Structure Chart, and suggest some ways in which some can be improved. In much the same way that we have expanded the original five levels of coupling [5] to seven, we will also expand the concept of coupling. We will include relationships other than calling module and called module. This enables us to address the kinds of issues and relationships present on real-time systems efforts. Real-time systems require relationships of a non-calling nature in order to conserve execution time. Synchronization and Informational Coupling account for these relationships.

12.3 SYNCHRONIZATION COUPLING

One of the most difficult problems facing designers of operating systems and other real-time systems is the issue of coordination among modules. The use of a simple passing of control from a BOSS module to each of several subordinate modules is expensive with respect to execution time in two ways.

First, transfers of control usually cost execution time because they can occur very frequently. In some multiprocessor systems, the occurrence of a CALL and RETURN will result in a switch in context or change of operating environment from one processor to another. This can cost a great deal of time, sometimes on the order of milliseconds.

The second problem with transfers of control may be the more significant of the two. It is that the advantage of a multiprocessor environment is greatly diminished. Processing occurs in sequence rather than in parallel. The concept of more than one module executing in the same logical time is mitigated.

Synchronization Coupling, as a concept and accompanying notation, is a direct result of a desire to apply Structured Design concepts to operating system developments by tailoring them to suit a problem. Synchronization Coupling occurs whenever two or more modules are coordinated by means other than a CALL/RE-TURN process. The exact means by which this may be accomplished is a function of the hardware, operating system, and programming language involved. In the design phase, it is necessary only for us to document that such a coordination arrangement will be used, not how it will be accomplished.

Figure 12.3-1 presents an example of the use of Synchronization Coupling. We will use the SUSPEND/RESUME concept (see Chapter 9) to explain the example. In the example, module C will SUSPEND itself unless module A and module B have issued RESUME commands to C. Such could actually appear in the pseudocode for the appropriate modules. There are usually many different ways in which synchronization can be described. The dictionary entry for the synchronization module might look like Figure 12.3-2.

Note that the synchronization module is not an executable, programmed module. Instead, it is used as a placeholder for us to document that some synchronization arrangement has been created between or among the modules indicated. Each

LABEL: SYNCH007

AUTHOR: J. BOND

DATE CREATED: 25FEB87   DATE LAST CHANGED: 27FEB87

MODULE DEPENDING   MODULE CONTROLLING

C   A

                                                      B

TIME LIMIT: 5 MILLISECONDS

TIMEOUT ACTION: RECORD CONTENTS OF REGISTERS A THROUGH F
                       ISSUE RESTART

            IF C REACHES DECISION POINT AND A AND B HAVE NOT BOTH
            ISSUED A RESUME C THEN

                        MODULE C ISSUES A SUSPEND C

            ENDIF

            IF C IS SUSPENDED THEN

                        IF A AND B EACH ISSUES/HAVE ISSUED RESUME C THEN

                                MODULE C IS RESUMED

                        ENDIF

            ENDIF

Figure 12.3-1: Generic Example of Dictionary Description of Synchronization

synchronization module has a unique identifier (e.g. Snnn, where nnn is an integer). Note that policy is described regarding the relationship between the modules involved, and since the suspension may last, what will be done if this time is exceeded.

One reason why this type of coupling is ranked at the top of our desirability list is that there is a minimum amount of interaction between or among the related modules. To test this out, consider what would be involved if module C were to be

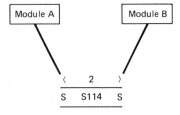

The above indicates that Module A SUSPENDs to syncbox S114, Module B issues a RESUME to S114 at some point in time. Syncbox S114 is described in the dictionary that there are 2 copies of this arrangement existing in the same logical time.

Figure 12.3-2: An Example of Synchronization Coupling

replaced. Neither module A nor B would be aware of it. No information passes so we do not need to concern ourselves with the possibility of a data-type mismatch. The utility of the Synchronization Couple may be limited to certain real-time situations but it does fill a very real need.

12.4 INFORMATIONAL COUPLING

This concept is similar to synchronization coupling, however the main difference is that information is passed between the synchronizing or contending modules. This information is data—not flags. Such messages can be earmarked for one of several competing tasks. This design technique also performs the handshaking function that occurs in a Procedure call. Another feature of this technique is that the module that receives the message may request it before it is sent, and still get it.

This type of coupling is not ranked lower on our scale because it is a rapid way to transfer information without incurring the overhead associated with a CALL and RETURN sequence or other less desirable ways of transferring information. Its use is restricted to real-time systems. An example of Informational Coupling is presented in Figure 12.4-1.

We have placed this coupling technique higher than others because either module could be replaced without causing a change in the other. As in the case of Synchronization Coupling (and the others) it is not necessary for us to define how this type of coupling can be accomplished but rather document the need for such an arrangement.

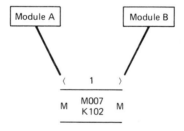

The above indicates Modules A and B are synchronized. M007 is the number by which this mailbox is referred to in the dictionary. K102 is the identifier of the lock/unlock parameter also in the dictionary. This arrangement has only 1 occurrence (the default).

Figure 12.4-1:   An Example of Mailboxing

12.5 DATA COUPLING

Data Coupling occurs when the information which passes between a calling and called module is unstructured. The character of this data is that it must be elemental. For example, in Figure 12.5-1, Modules X and COMPUTE_SINE are Data Coupled. The reason why Data Coupling is the most desirable of the possible module

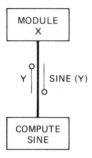

Figure **12.5-1:** An Example of Data Coupling

interactions is that it is the most explicit. No "knowledge" regarding the nature or organization of the data is needed by either module. Interfaces of this type are explicit in that nothing needs to be assumed in order for the interface to work. In a sense, what we are trying to do is create X-rated software designs, since nothing is left to the imagination. Simple interfaces of this type make it possible, even easy, to replace one module with another without the software engineer having to know anything more than what data is passed and what function the replacement module must perform. Systems which lack this "functional pluggability" tend to be expensive, if not impossible, to maintain.

A rather insidious variation on Data Coupling is "wideband" transmission or "swarming." In this instance, a number of data couples are being passed. If so many data couples are needed, then the character or cohesion (see Chapter 13) of one or both modules is suspect. An example of "swarming" is presented in Figure 12.5-2. It is sometimes used by designers who do not wish to admit that Stamp Coupling (see Section 12.6) is what they are really after.

Data Coupling is the highest level of coupling that can be used in any type of system. It may not always be the best way to communicate information. For example, high data volumes or high rates of data transfer would consume valuable resources due to the passing of information through parameter lists.

The advisory regarding this and any other form of coupling is to use what makes sense under the circumstances. There is not a "best" form of coupling except within the context of the operational constraints within which it is being used.

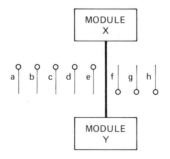

Figure **12.5-2:** An Example of a "Swarm"—Not Recommended

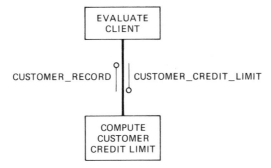

Figure 12.6-1:   An Example of Stamp Coupling

12.6 STAMP COUPLING

Some people have complained about the name ''Stamp Coupling'' because it does not seem to communicate the nature of this type of relationship between modules. Philatelists may disagree, since a stamp is, in a sense, a pattern which is agreed upon by sending and receiving parties. Over the last seven years, complainants have been asked to suggest a better name. To date, no better one has been suggested.

Stamp Coupling occurs when two modules pass data which have some structural characteristics. An example of Stamp Coupling is presented in Figure 12.6-1. The modules EVALUATE_CLIENT and COMPUTE_CUSTOMER_CREDIT_ LIMIT are Stamp Coupled because the data that passes between them has a structure. That is, CUSTOMER_RECORD is composed of several elements, and these are organized according to a specific layout. Although CUSTOMER_RECORD is what is being passed, little or nothing is done with it as an entity. What are used are the *elements* within CUSTOMER_RECORD. In order to make this possible, the module COMPUTE_CUSTOMER_CREDIT_LIMIT and *any other module* using elements in CUSTOMER_RECORD need to know how CUSTOMER_RECORD is organized. This shared knowledge is, in a sense, dangerous. Replacing either mod-

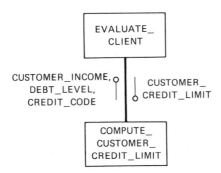

Figure 12.6-2:   Modification of Stamp Coupling to Data Coupling

ule involves knowing both the function to be performed by that module and the organization of the Stamp Couple. This shared knowledge is what causes us to view this type of relationship as being less desirable or more dependent than Data Coupling. In general, the module receiving such structured data only uses a subset of the data items—not all of them. An improved situation is presented in Figure 12.6-2. Note that only the parameters used are being passed and we have changed the level of coupling to Data Coupling.

12.7 CONTROL COUPLING

Control Coupling occurs when a flag passes between two modules. In this text, "flag" means information which resulted solely from our implementation of this system. That is, flags are *not* inherent in the problem or system studied in the analysis. They are an inherent part of the *solution*. Such information is used by one module to signal another that a certain condition has occurred (e.g., end-of-file, memory overflow, invalid data). It is not information which is inherent in the enterprise which was analyzed during the Structured Analysis phase. The information passing from module EVALUATE_IFF_RETURN and module CLASSIFY_TARGET in Figure 12.7-1 constitutes a Control Couple. This type of relationship is considered more complex (less desirable) than the others we have discussed because it requires "operational knowledge" of one module to be possessed by the other module in the relationship. This is so because flags have the effect of signalling a condition, they have a limited range of values, and they cause the called module to behave differently from call to call depending on the value of the flag received. More about this point when we discuss cohesion.

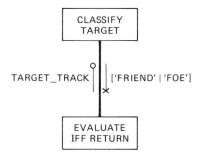

Figure 12.7-1: An Example of Control Coupling

12.8 COMMON COUPLING

Most of us have seen movies where a spy puts the microfilm or other secret document into a newspaper, leaves it on a park bench, and walks away. Later, another spy casually sits down at the bench and eventually leaves with the newspaper, microfilm and all. Although this often works in the movies, problems can be encountered. The assumption on the part of both spies is that no pigeon will ruin the microfilm and no vagrant will walk off with the paper. Any programmer who has used Common Coupling knows that there are lots of pigeons around. This "spy drop" approach describes Common Coupling.

Common Coupling occurs when data is transferred from one module to another via a common storage area. Examples of such COMMON areas are presented in Figure 12.8-1. The main reason for ranking this type of coupling lower than the others discussed so far is that this interface is highly implicit. It is more difficult to deal with in maintenance and debug modes. Implicit interfaces of this sort work only when there is mutual understanding among the software designers involved. As an example of what can happen when this type of interface is employed without restraint, a system of over 200 FORTRAN routines was to be converted from a UNIVAC system to IBM. Over 20 labeled COMMON areas were used to pass data from one routine to another. Almost half the conversion effort was devoted to tracking down problems caused by one routine or another which "clobbered" some portion of a COMMON block, leaving no clue as to which one "did the dirty deed." By means of core dumps and other schemes, such problems were eventually removed. The point is, when COMMON is used, the software engineer has very little evidence with which to determine where and how portions of it may have been tampered with. A "bug" in one module which interfaces with the COMMON area may cause widespread, uncontrolled problems, the cause of which will be difficult to identify.

In some situations the use of Common Coupling appears to be inevitable. These include many real-time applications such as simulations, data-gathering instrumentation systems, and operating systems. These types of systems are popular places to use Common Coupling because the volume of data, the number of parameters, the operating system, the hardware interface, or some combination of these will consume an unacceptably large amount of time if data is passed through a parameter list. But there are some simple ways to "pull the teeth" of the Common Coupling monster.

Note that there is more than one problem with Common Coupling. The difficulty of finding out which module ruined the data in the COMMON area is only one of them. Another, which can easily occur in COMMONs where there are many parameters, is misnaming or misordering the parameters. This will cause problems

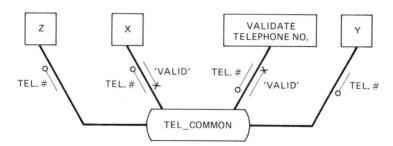

Figure 12.8-1:   An Example of Common Coupling

even if the rest of the code is "error free." The temptation is to just use Common Coupling and ignore the potential for problems.

An improved, hybrid form of COMMON is supported on several systems which are used for real-time applications. This technique involves a specialized type of comment within each module that sets/uses one or more COMMON parameters. It works this way: each module has a header or lead-in block of special comments. For example, in FORTRAN such a comment might look like:

```
C* ALT_KM
```

A preprocessor examines each module, creates a correct labeled COMMON block for modules with such specialized comments, and sends this "new" set of code to the compiler. The order in which the comments are presented to the preprocessor is not important, since it will arrange the executable part of the COMMON correctly. This is not an ideal situation, but it is an example of how a reasonable compromise can occur between the ideal (i.e., avoid the use of COMMON) and reality (i.e., other approaches in that environment are not viable). Some programming languages accomplish the same thing using special constructs.

12.9 CONTENT COUPLING

Content Coupling occurs whenever one module uses or modifies code in another module. Two common ways of doing this are:

Use of the ALTER feature in COBOL

One module branching into another, executing several instructions within that module, and returning control to itself

This type of relationship is ranked at the bottom because it is the most insidious of the lot. Some software departments feel so strongly about the danger of this type of practice that they have literally outlawed it. By now, the reason for this should be obvious. The person who is attempting to understand the system well enough to modify it is at a serious disadvantage. Seemingly innocent changes to a module will produce unexpected, unpredictable, and undesirable side effects.

The repercussions of such practices were clearly evident, for example, on a satellite data reduction system. A "clever" programmer had a "knack" for coming in under the memory budget for the functions he was assigned by deftly utilizing code in other people's modules. He did this by exiting his own module at some point, branching into someone else's module at a certain point, executing a number of instructions, then reentering his own module at the next instruction. It worked well until he left the company. Prior to that time, he kept well apprised of changes in these other modules. After his departure "weird" system crashes occurred. A change to some module would cause the system to fail at a location and within a

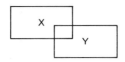

Figure 12.9-1: An Example of Content Coupling

function that simply did not make sense. It did not make sense because it was an unrelated task. Diligent efforts revealed the ''modus operandi'' of this individual, and changes were instituted. During the interim, a great deal of human resource was expended trying to explain the impossible.

Coding and design practices of this type have been widely used in certain real-time applications owing to a misguided understanding of the word ''efficiency.'' Often, the argument used to justify such practice is that storage space is limited and, hence, code sharing is inevitable. Although it may be true that storage space is limited, analysis of such situations has shown that most often, those involved had not done an adequate job of partitioning. As a result, certain functions were duplicated throughout the system. A better, more detailed job of partitioning would have revealed these, enabling the software engineers involved to eliminate duplication with a resulting reduction in storage requirements. An example of what Content Coupling would look like graphically is presented in Figure 12.9-1.

12.10 EVALUATING COUPLING

Coupling between modules can be evaluated by identifying what data passes directly between them. That is, the modules involved must be hierarchically related in a Caller/Callee relationship. Classifying the nature of this data can be a little difficult. A summary of classification guidelines is presented in Table 12.10-1.

TABLE 12.10-1: Summary of Classification Guidelines

Type of Information	Type of Coupling
Non-hierarchical coordination (e.g. SUSPEND, RESUME)	SYNCHRONIZATION
Restrictive, asynchronous message passing (e.g. MAILBOX)	INFORMATIONAL
Unstructured data (e.g., data elements)	DATA
Structured data (e.g., a record)	STAMP
Condition signal (e.g., an end-of-file flag)	CONTROL
Data widely available for use and change (e.g., common block)	COMMON
Shared or modified instructions within a module (e.g., COBOL ALTER)	CONTENT

These guidelines are designed to enable the software design engineer to assess the level of coherence present in the system via evaluation of pairs of modules. By *coherence* is meant the level of togetherness or synchronized interaction exhibited by the system. The simpler the interactions, the greater the degree of coherent interaction. A similar situation exists in the field of laser technology. Incoherent light, such as we encounter in ordinary artificial light and sunlight, is composed of light waves whose wavefronts are randomly arranged. They are, basically, harmless—not powerful. However, the mere expedient of synchronizing those wavefronts using a laser device has the effect of creating an extremely powerful and potentially dangerous burst of energy.

12.11 IMPROVING COUPLING LEVELS

Now that we know how to evaluate the level of coupling exhibited by a module, the next step is to identify what can be done to improve the coupling of certain module pairs. Table 12.11-1 lists coupling levels and some suggested ways of improving them.

TABLE 12.11-1:   Coupling Levels and Improvement Schemes

TYPE OF COUPLING	TO IMPROVE LEVEL
Synchronization	Ensure that this is really needed.
Informational	Ensure that this is really needed.
Data/"swarm"	Check to make sure that all elements are really needed. May actually be a case of Stamp Coupling. Replace "swarm" with a data structure if elements are actually related. Otherwise split one or both modules up into several more functional modules.
Stamp	Have modules exchange *only* the data elements needed.
Control	Sometimes unavoidable. Make sure it is in this case. Keep the number of Control Couples low between any two modules.
Common	Move the modules which produced the needed data closer to the module which uses it—usually in a subordinate/superordinate role, respectively.
Content	Send the design's author to a home for the terminally weird. Try to undo the harm left behind.

12.12 THE CONCEPT OF COHESION

Coupling enables us to evaluate the quality of the relationship between modules—but what of the quality of each module? *Cohesion* addresses this issue. Under this concept, a module is viewed as having desirable characteristics if it is highly functional. In this case, "functional" can be thought of as the degree to which all the elements or instructions which comprise the module contribute to the accomplishment of a single task. This task need not necessarily be a simple one. If it is overly complex, the module could be broken up into a few simpler, more functional ones.

Cohesion can be thought of as the other side of coupling. It is, in a sense, a microscopic view of a system, where the smallest object that can be resolved is an instruction. Similarly, coupling is a macroscopic viewpoint, where the resolution limit is the module. As in the case of coupling, Structured Design provides us with a categorization of the cohesion levels. There are seven categories, ranked from most desirable to least desirable.

Modules which exhibit the more desirable cohesion levels are composed of elements exhibiting the highest coherence. That is, much like the wavefronts of light in a laser, they are all in step. They are focused on a single end, goal, or function. Less desirable modules lack this coherence and tend to have unclear or multiple goals, none of which are clear-cut. In a sense, such modules tend to dissipate their energy, reducing their positive impact. More important, such dissipated modules present maintenance programmers with a real intellectual challenge. Via a personal survey, maintenance programmers have indicated that their least favorite type of module to have to deal with is the ''do everything'' module—the type that receives a flag telling it which of its many services is desired this time. The required service(s) may well be different the next time it is called. In many important ways, modules with poor cohesion suffer badly from an identity crisis.

We will discuss cohesion in more detail in Chapter 13.

12.13 COUPLING AND COHESION—A SUMMARY

Together, these two types of evaluation criteria form a system for judging the relative quality of a software system design. Since design does involve judgment and experience, situations may arise where it is difficult to discern precisely what level of coupling or cohesion is appropriate to assign. The thing to keep in mind is that we are not dealing with an absolute situation but with one which is relative. It is a misconception to believe that the *exact* level of coupling and cohesion will always be differentiable. The key to getting the maximum benefit from the use of the coupling and cohesion concepts is the consistent, disciplined application of the evaluation criteria. In this way, the relative quality of a module or intermodule relationship can be established, since we are not dealing with some absolute scale.

Another misconception regarding the use of Structured Design is that the coupling concept and cohesion concept can be used only if Structured Design is the design method used. The facts of the matter are that both coupling and cohesion can and should be used with other methods. After all, relationships between modules are simple, complex, or something in between regardless of the method (if any) by which the system architecture, module population, and intermodule relationships were established. Viewing coupling as being exclusively part of Structured Design is simply incorrect. Hence, these two concepts in Structured Design are transportable not only among systems but among methods as well.

The idea that Structured Design concepts are useful for design but not for maintenance is another fallacy. In order to maintain a system, we need to understand it. What better way to understand it than to identify and evaluate the nature of the relationship between modules and the relative quality of each of the individual modules? Coupling and cohesion can be an integral part of a plan for system improvement. By identifying those modules and relationships that are markedly undesirable in comparison with the rest of the system, we can target certain modules or groups of modules for improvement. Consider this: without such a strategy, current systems which have some degree of usefulness left in them will probably outgrow that level of usefulness in a relatively short time. What is to become of the investment that was made in developing and maintaining such systems? Conversations with maintenance programmers at several major corporations indicate that the cost of replacing any of the software systems vital to their employers is prohibitive. That is, the company could not afford to replace them without stressing itself financially. Hence, the firm continues to keep many of its systems "limping along" via the expenditure of vast amounts of human resources. The people involved in such situations can get pretty testy once they realize that, in spite of their recommendations, the firm declines to replace this blatantly unmaintainable system.

The application of coupling and cohesion as part of an overall strategy aimed at system improvement offers one very effective way out of this morass. Without such an objective value system with which to measure progress and identify and prioritize the areas of the system most in need of improvement, the system upgrade effort gets diluted and loses its impact. Also, the use of these quality criteria provides maintenance people for the first time with a means of measuring the relative costs of various types of system and module construction practices. If one contemplates the enormous amount of software in use within corporations today, there is only one rational conclusion regarding the replacement of software systems: it is less than unlikely that any large number of them will be replaced, owing to the great expense involved. Hence, many new concepts will have to be employed in the area of software maintenance. The quality criteria employed by Structured Design seem likely candidates for adoption into this next phase of software engineering practice.

REFERENCES

1. T. J. Peters and R. H. Waterman, Jr., *In Search of Excellence—Lessons for America's Best-Run Companies.* New York: Harper and Row, 1982.

2. H. W. J. Rittel and M. M. Webber, "Dilemmas in a General Theory of Planning," Institute of Urban and Regional Development, Working Paper No. 194, Berkeley, University of California, November 1972.

3. L. J. Peters and L. L. Tripp, "Is Software Design Wicked?" *DATAMATION*, Vol. 22, No. 5 (May 1976), pp. 127–36.

4. W. P. Stevens, G. J. Myers, and L. L. Constantine, "Structured Design," *IBM Systems Journal,* Vol. 14, No. 2 (May 1974), pp. 115–49.

CHAPTER 13

Evaluating and Improving Modules

In the previous chapter we discussed the module-level view at which Structured Design analyzes systems. The smallest unit we examined was the module. Emphasis was on interactions between modules. What went on within modules remained something of mystery. In this chapter we will discuss another, more detailed view utilized by Structured Design. The smallest unit examined at this level is the individual instruction; the largest unit, the module. We examine each module on its own merits without regard to how it relates to other modules in the software system.

The need for this more detailed view can be demonstrated by using an analogy. Other types of systems, such as bridges, rely for stability on both the quality of their components and the arrangement of those components. Faulty components or materials can cause the bridge to fail. Similarly, even if the components are sound, a poor arrangement of components can cause failure. Hence, stable systems need both quality components *and* a prudent arrangement of them.

Most of the characteristics of a quality design may already be intuitively obvious to even the novice software engineer. But in the area of module quality, as in the area of system quality, Structured Design employs a refined and quantified evaluation of alternative designs to form a discipline. The term *cohesion* refers to the character of an inherent property of a given module—*not* a property which is either present or absent. The main difference between modules is that some are more cohesive than others. That is, the character of a module has certain classifiable properties.

This chapter describes the concept of cohesion, how it is incorporated into

Structured Design, how to evaluate the cohesion of a given module, and what the levels of cohesion are. The reader should remain aware that nothing precludes the use of this concept in other software applications such as maintenance. The greatest use will be made of this concept if the reader adopts it and employs it as a means of measuring one module design alternative against another in an objective way. Remember that these alternatives occur not just during design, but during implementation and maintenance as well. Cohesion represents one of the two primary conceptual tools Structured Design offers us which can be applied to any software design or system.

13.1 THE CONCEPT OF COHESION

Historically, the problems associated with the timing and sizing of software systems led to a philosophy of trying to keep the overall number of modules low. One result was modules which had multiple capabilities. Each module could be used for many purposes. Unfortunately, such modules became intellectually intriguing, and interfaces and protocols between modules became complex. Today, we are reaping the harvest of such practices. Few, if any, software engineers thought that the systems that they developed in the late sixties and early seventies would still be in use today. Some systems developed even earlier are in use at major corporations and government agencies. The maintenance problems caused by such systems have prompted a reexamination of these philosophies and resulted in many recommended courses of action during the software design phase. For example, many now realize that simple modules, related in simple ways, are easier to develop and maintain.

It may be difficult to change habits and views, but little by little, software engineers are beginning to accept the ''simpler is better'' philosophy. This makes it easier to maintain the software and also reduces the possibility that the engineer will be snared by the ''maintenance potential well.'' This phenomenon results when only one engineer understands a particular system. Management is usually reluctant to let that engineer transfer to another organization within the firm, since there is no replacement who has the same specialized knowledge. In a sense, that software engineer is trapped by the system being maintained, much as a planet is trapped in its orbit about the sun. The only way out is to leave the firm altogether.

Even given the acceptance of a more enlightened approach, the need remains for a means of measuring or evaluating how *simple* a module is. Cohesion provides us with a measurement mechanism. It is to measure the level of coherence exhibited.

13.2 LEVELS OF COHESION

When we discussed coupling, the degree of simplicity formed the basis for classifying relationships between modules. In the case of cohesion, simplicity is important, but the degree of specialization is also considered. Up to a point, the more specific a function that a module performs, the more desirable it is from a cohesion standpoint. It is also more desirable from a reuse standpoint. General-purpose, "do everything" modules are usually among the "hard core" unreusables in a system. Overly specialized modules, particularly ones that suffer from the use of "magic numbers" [1], also fall into this category. What we seek are modules which may be of use in some other system or even on an entirely different application.

We have identified seven different levels of cohesion. These will be discussed in order from most desirable to least desirable.

13.2.1 Functional Cohesion

The most desirable level of cohesion is called *functional* because every element (i.e., instruction) that comprises a functionally cohesive module is directed at accomplishing a specific, discernible task. It may not be a simple task, but it is just one task. Examples of functionally cohesive module names include Validate Account Number, Compute Penalty, Compute Sine of X, and Sort Receivables by Amount. This type of module may seem somewhat dull, but it comprises the backbone of a well-designed system. One positive aspect of functional modules is that there is very little to understand about them. As a result, systems populated with them are more quickly and easily understood by people who did not author them. The basic rule about distinguishing functional modules from others is this: if it does one, well-defined job, it is functional. This type of module possesses a high degree of reusability because of its utilitarian nature.

Our definition raises some serious questions regarding the complexity of functional modules. For example, what if the function is TRACK_TARGET? This can be pretty complicated. It will involve gathering raw data points, validating raw data points, checking to see if they fall within an existing track, and, if they do, checking to see if we now have a qualifying track. If so, we form a track box, etc. Not so simple, is it? With all this going on, could TRACK_TARGET be a *functionally* cohesive module? Yes, if it is the boss of subordinate tasks. Otherwise, it needs to be split up.

13.2.2 Sequential Cohesion

A module is sequentially cohesive if the elements which comprise it exhibit a production-line interdependence. That is, each element, in turn, uses the output of the preceding or previous element as its input, does something to it, and sends its output on to the next element. That element's input is the previous element's output, and so on. This type of module is highly cohesive. Its sequential or multiple ordered-step nature makes it somewhat less desirable than a functionally cohesive module. Note that it is the order of the process *and* the interdependence of data that links the elements of sequential modules. Pseudocode for a functionally cohesive module might look something like that in Figure 13.2.2-1.

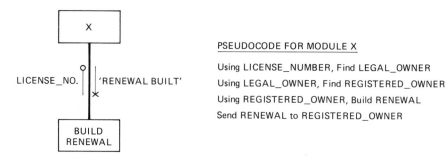

PSEUDOCODE FOR MODULE X

Using LICENSE_NUMBER, Find LEGAL_OWNER
Using LEGAL_OWNER, Find REGISTERED_OWNER
Using REGISTERED_OWNER, Build RENEWAL
Send RENEWAL to REGISTERED_OWNER

Figure 13.2.2-1:   An Example of Sequential Cohesion

13.2.3 Communicational Cohesion

A module is communicationally cohesive when its elements all use the same (input or output) data but the results of one stage in the process are not passed on to other stages. What binds together the elements in communicationally cohesive modules is the data they use. This is a weaker type of bond than the function or process. An example of a communicationally cohesive module in terms of pseudocode is presented in Figure 13.2.3-1.

13.2.4 Procedural Cohesion

Procedurally cohesive modules are composed of elements which are related by the order of their occurrence. This type of module is often confused with sequentially cohesive modules, but there is an important difference. Sequentially cohesive module elements are linked by element order *and* the data fed from one element to another. Procedurally cohesive module elements are linked strictly by the order in which they are to be executed. Data does not feed from one element to another. An example is presented in Figure 13.2.4-1.

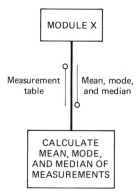

Figure 13.2.3-1:   An Example of Communicational Cohesion

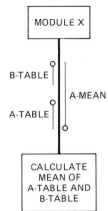

Figure 13.2.4-1: An Example of Procedural Cohesion

13.2.5 Temporal Cohesion

Modules whose elements are linked by the time at which they are to be executed are said to be temporally cohesive. The most common modules in this class are initiation and termination modules. Such modules often result from a "lockstep" approach to system startup and shutdown. Consider this aspect when examining a temporally cohesive module which, for example, initializes all flags, pointers, etc.: do they *all* have to be initialized right then, or is this necessary only for some, while the others could be forestalled until another phase of the processing? An example is presented in Figure 13.2.5-1.

13.2.6 Logical Cohesion

Logically cohesive modules are composed of elements performing tasks which can be classed under a single, broad category. Modules of this type could be renamed "do everything" modules. Because of their multifaceted nature, these modules are usually called and sent one or more flags. The flag tells the module just what sort of service it is to perform *this time*. It may well perform a different service the next time it is called. In a manner of speaking, such modules suffer from an identity crisis.

We are not rating logical cohesion sixth in a field of seven because we are

SET TOTALS TO ZERO

OPEN FILES

SET PAGE # TO 1

SET LINE # TO 1

Figure 13.2.5-1: An Example of Temporal Cohesion

PRINT HEADING

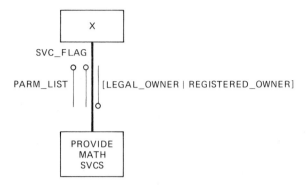

PSEUDOCODE FOR PROVIDE CAR DETAILS

Using PARM_LIST & Svc_flag
If Svc_flag = 1
    return class
If Svc_flag = 2
    find legal owner
    return legal_owner
If Svc_flag = 3
    find registration
    return reg_owner

Figure 13.2.6-1:   An Example of Logical Cohesion

being discriminatory toward modules with psychological disturbances; rather, we are trying to avoid the side effects that they produce. Specifically, these weakly related modules usually contain some amount of code sharing. Instead of having a distinct set of elements for each type of service, a programmer will notice a similarity between certain parts of one code segment and another. Often, he or she will use the "common" code for more than one type of request. The observation, "It's a jungle out there," must have been invented by someone who had to maintain such modules. An example is presented in Figure 13.2.6-1.

13.2.7 Coincidental Cohesion

In this case, there is no apparent bond between the elements. They are in the module because, perhaps, there was noplace else to put them. There is basically no excuse for allowing the continued existence of such "garbage can" modules. Figure 13.2.7-1 provides an example.

```
CALCULATE FICA DEDUCTION
            OR
COMPUTE POINT OF MISSILE IMPACT
            OR
CALCULATE INVENTORY REORDER AMOUNT
```

Figure 13.2.7-1:   An Example of Coincidental Cohesion

13.3 EVALUATING COHESION

The procedure used to evaluate cohesion is a simple one:

1. Write a sentence that describes what the module does.
2. Analyze the sentence.
3. Classify the module's cohesion according to the preceding set of guidelines.

Although step 1 is simple, the analysis required to do steps 2 and 3 is not. Note that when Structured Analysis is used as the specification approach, step 1 may already be taken care of for many modules in the form of structured English. In either case, analysis of the descriptive sentence is performed as follows:

If the module description can be summarized via a strong-verb, object-noun combination and is clearly related to a task within the scope of the system, then it exhibits *functional* cohesion.

If the module description involves an ordered set of subtasks, the output of each feeding the next as input, then it exhibits *sequential* cohesion.

If the module subtasks do not relate sequentially and they utilize the same input data, then it exhibits *communicational* cohesion.

If the module subtasks are ordered but do not ''feed'' each other (as in sequential cohesion) and do not all work on the same input data (as in communicational cohesion), then it exhibits *procedural* cohesion.

If the sequence of the module's subtasks is not relevant but time is a consideration (e.g., initialization, termination), then it exhibits *temporal* cohesion.

If the module can perform any of several related subtasks (the desired subtask being selected via some sort of flag or trigger), then it exhibits *logical* cohesion.

If the module can perform any of several subtasks but they do not appear to have any apparent relationship, then it exhibits *coincidental* cohesion.

13.4 SUMMARY OF COHESION LEVELS

We have described the seven levels of cohesion in descending order from most desirable to least desirable. Their status, or acceptability, is summarized in Table 13.4-1.

The desirability of one level of cohesion or another is relative in two ways. First, it is relative to other levels of cohesion. That is, there is a desirability ''pecking order'' of sorts which ranks these levels from best to worst. Second, cohesion levels vary in their applicability within a specific environment. That is, functionally cohesive modules may *not*, necessarily, be the best way in which to

TABLE 13.4-1:   Cohesion Levels and Their Desirability

COHESION LEVEL	STATUS
Functional	Most desirable
Sequential	Acceptable
Communicational	Acceptable
Procedural	Occasionally useful
Temporal	Occasionally useful
Logical	Should be avoided
Coincidental	Outlaw

implement a system. This is not quite as heretical as it may sound! The programming language, operating system, and hardware characteristics in which the system will run *must* be considered during physical design. For example, some operating environments experience serious timing problems when several hierarchical calling levels are employed. More about this later.

13.5 IMPROVING COHESION LEVELS

Often, though not always, cohesion levels may be improved. Care should be taken to assess the ramifications of your proposed "improvement" to make sure that matters have, in fact, improved. A list of levels and possible improvements is presented in Figure 13.5-1.

COHESION LEVEL	POSSIBLE IMPROVEMENT
Functional	Great! But watch out for overly simple modules during the physical design phase.
Sequential	If the individual stages are complex or long enough, consider repartitioning into several modules.
Communicational	Repartition into several simpler, single-task modules.
Procedural	Repartition into several simpler, single-task modules.
Temporal	Often unavoidable. Be sure that any initialization(s) occur as "late" as possible during processing. Avoid the large, "big bag" oriented temporal module.
Logical	Separate individual functions into several single-function modules.
Coincidental	Contact a toxic waste dump, then break module up into simple, more functional modules.

Figure 13.5-1:   Cohesion Levels and Improvements

13.6 STRUCTURED DESIGN AND COMMON SENSE

At first, software designers may resolve that from now on, only functionally cohesive modules will populate their designs.

> WARNING: The author has determined that resolving to create designs populated only by functional modules may be hazardous to your professional and mental health!

Possibly the worst, most imprudently implemented systems this author has ever seen were the result of such a mind-numbing approach to software design. Totally populated by functional modules, some of them as small as three executable lines, these systems clearly demonstrated what can happen when one substitutes rules for thought. Designers lost sight of the fact that these are advisories or guidelines, *not* absolute rules. There may be perfectly good reasons for employing just about any level of cohesion (or, for that matter, coupling) in a system. Yes, that may sound like heresy, but it is true. The goal of the designer is still to develop the architecture for a system which meets the client's needs in a cost-effective way within the constraints of time and money as imposed by the project. The sole purpose of the evaluation schemes utilized by Structured Design is to give the software design engineer an indication of the *relative* quality of certain aspects of the design and not of right or wrong design practice. Design is, first and foremost, an amoral activity—any means to the goal of a working design is permissible.

Surely, some readers will interpret this to mean that they can do whatever they please and this coupling and cohesion stuff is just so much structured hype. If that is the case, then the point has been missed. The point is that Structured Design (and Structured Analysis) is an effective adjunct for our reasoning abilities and judgment and *not* a replacement for them.

The level of coupling or cohesion that is present in a system or within a given module is, to a certain extent, not the issue. What is at issue is whether or not the software designer has considered alternatives. That is, of all the possible ways of defining and arranging the modules that comprise the system and defining their inner workings, was there a better way which was rejected? The actual levels of coupling and cohesion present are not the important thing. What is important is that the levels that were decided upon were the result of considering and rejecting several alternatives based on a value system. A major merit of Structured Design that has gone almost unnoticed by most authors is that it provides a common, defined, externalized value system. This value system can be used to assess the quality of an existing system, the quality of a proposed design, and the relative quality of alternative design architectures.

In a manner of speaking, that is what design is all about—selection from among alternatives. However, in order to do this we must have a basis or value system by which one alternative or another may be accepted or rejected. We are not hereby suggesting that poor-quality levels of coupling or cohesion are OK. Quite the contrary. What is being suggested is that justification of most levels of coupling and cohesion is possible *only* if many alternatives have been proposed, a value system uniformly applied, and a selection made.

13.7 ADDITIONAL MEANS OF EVALUATING A MODULE

The concept of cohesion deals with the *apparent* level of coherence present in a module. That is, it addresses what the module appears to be like based on a description of what it does. This makes functionality the benchmark for individual module quality. Hence, a module whose description is "This module validates authorization codes" may be declared to be functionally cohesive. If we devised a more detailed description which went into just how this validation was done, we might wish to alter the level to sequential cohesion or some other.

Regardless, the fact remains that this approach puts a lot of faith in the designer/developer's ability to carry out the task of implementing this "module specification" in a rational and reasonably coherent manner. Structured Design in its original form does not provide us with a means of establishing that such implementation actually took place. That is, under the cohesion approach, we *assume* that if a module's cohesion is functional or procedural or whatever, all is well. In fact, all is not well! In numerous cases the functionally cohesive module has been implemented in a gross manner. It was not executed into code in anything like the way one might have assumed. Perhaps the programmer had a "bad day," was coming down with a cold, or whatever. The cause is not important. The effect is. The fact is that Structured Design in its original form does not provide any means of determining that what was so well thought out through the structured English development remained well thought out through pseudocode (and code) development.

Once again, we are dealing with a three-faceted system of evaluation. One facet, coupling, relates to the nature of the relationship(s) which exist between and among modules. Another, cohesion, relates to the internal coherence of each module. The third, complexity, relates to the degree to which this coherence was instantiated into a pseudocode or code expression of the module.

Several different models of complexity are available today. Most were developed in an attempt to evaluate code for various properties. One of the simplest and easiest to use (actually, easiest to automate) is the GREENPRINTS model suggested by Les Belady [2]. In it, code or pseudocode which contains many nesting levels is given a higher complexity rating than a module which has fewer. Also, the greater the number of instructions at a nesting level, the greater the complexity. This effect is increased as a function of the depth of the nesting level at which the instructions appear. The complexity factor is computed from the formula:

$$C = \frac{S}{r_t} \sum_{r=1}^{r_t} N_r$$

where

    C = the value of the complexity factor

    S = the total number of lines of pseudocode or code for this module

    r = the nesting level of the rth line

    r_t = the total number of nesting levels

    N = the nesting level of row r

    To check out the validity of this complexity factor, we may wish to make some assumptions:

> Poorly written code and pseudocode will have a higher complexity factor than well-written code or pseudocode which implements the *same* algorithm.
>
> Rewriting poorly written code to improve its quality will reduce its complexity factor.

Another text [3] already discusses this. There, the examples presented in [1] were used. In [1], poor code was presented and improved. In *every* case, the improved code showed a lower complexity than the poor code. Similar results are obtainable and have been obtained with pseudocode.

13.8 COHESION—A SUMMARY

Cohesion is a valuable ally in our quest for quality modules and systems but it is a factor that we must consider together with others. The level of cohesion possible in most software design situations is nearly dictated to us. That is, certain types of modules would be inappropriate to a given situation. Other types of modules would not be able to get the function accomplished in a reasonable way. This leaves some subset of module types that should be considered. The software designer's task is to recognize what the range of module types is that would be appropriate and to select quickly from among those. Creating software design that are composed of one or two module types exclusively is clearly *not* the answer.

REFERENCES

1. B. W. Kernighan and P. J. Plauger, *The Elements of Programming Style*. New York: McGraw-Hill Book Company, 1974.
2. L. A. Belady, C. J. Evangelisti, and L. R. Power, "GREENPRINT: A Graphic Representation of Structured Programs," *IBM Systems Journal,* Vol. 19, No. 4 (1980), pp. 542–53.
3. L. J. Peters, *Software Design: Methods and Techniques*. New York: Yourdon Press, 1981.

CHAPTER 14

Other Design Heuristics

This chapter discusses many characteristics of quality software design which were not described in the previous chapters. Many of these have been adapted from other fields, and some are based on first-hand experience and a ''gestalt'' event of sorts. Some may be getting their first public airing in this text. These are techniques which the author has used in project management and consulting work and which have pointed out areas in need of further investigation on a smaller, more detailed scale.

It is not the intent of this chapter to undermine the techniques which were described earlier. Quite the contrary. These techniques are intended to help the reader to quickly identify whether or not there is a problem. The techniques described earlier will help to alleviate it.

In the previous chapters we discussed two primary means of evaluating and improving the nature of the interactions which occur between modules and the modules themselves. These are not the only ways in which the quality of a software system design can be evaluated. Other, less formal and detailed ways have been learned through experience. Most of these enable the designer to quickly identify problems and take remedial action *without* having to go through the sometimes tedious process of (for example) writing out a sentence and analyzing it. Many of the comments contained in this chapter are responses to requests from students and clients who have asked where they can get more information about how to spot this or that in a Structured Design.

Although there are many different types of heuristics (learned practices based on experience) which we could discuss, we will focus on a few that have proven to be especially useful. Six problem areas that we will examine relate to different aspects of design quality. They are:

Decision Splits
Tramp Data
Information Hiding
Qualitative Evaluation based on the Structure Chart
Functional Analysis
Relationality

Each of these approaches compresses or consolidates several aspects of Structured Design and software engineering practice into a simple rule of thumb. They indicate quality level as well as what remedial action should be taken.

14.1 DECISION SPLITS

A common problem that software designers build into their systems is the Decision Split. This occurs when a module which discovers a condition is widely separated in the Structure Chart from the module that must do something about the condition. Such a situation is depicted in Figure 14.1-1. Note how module J detects a problem

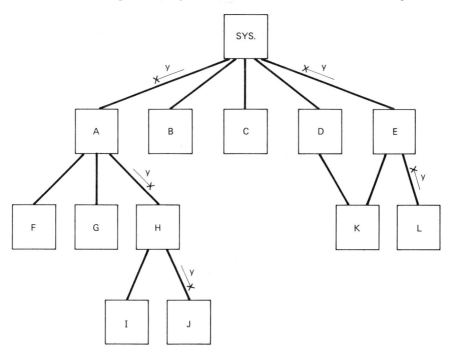

Figure 14.1-1:   An Example of a Decision Split

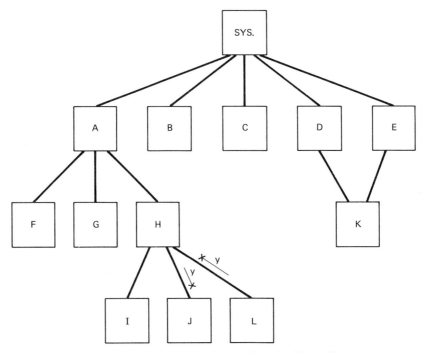

Figure 14.1-2:   One Way of Correcting a Decision Split

but module L is tasked to do something about it. To correct this, either move module L to be subordinate to module H (Figure 14.1-2) or move module J to be subordinate to module E. Either of these is a viable option, because discovering a problem and acting on a problem are closely related functionally.

14.2 TRAMP DATA

A phenomenon similar to the Decision Split is Tramp Data. The difference visually is that a data couple is being passed through the system instead of a control couple (Figure 14.2-1). The reason is that the module which uses a certain data item is widely separated from the module which produces it. As in the case of the Decision Split, we need to move the data producer closely to the data user (Figure 14.2-2).

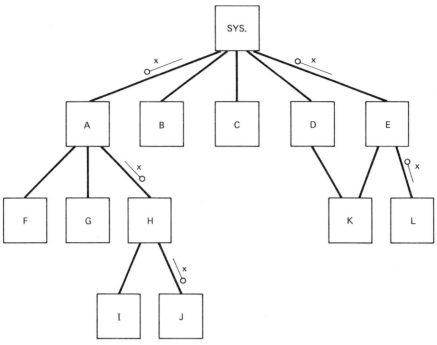

Figure 14.2-1: An Example of Tramp Data

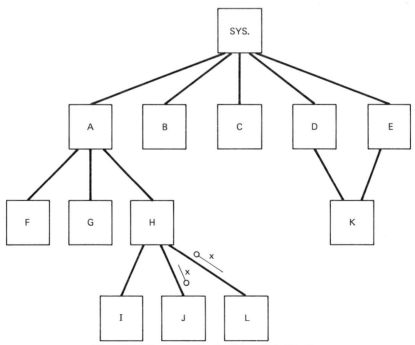

Figure 14.2-2: Improving the Tramp Data Situation

14.3 INFORMATION HIDING

There is something almost sinister about the sound of the name "Information Hiding." The intent of this technique is to restrict knowledge possessed by modules in the system on a "need to know" basis. It is an extension of the "starve your bubbles" concept in the analysis phase. In some respects, the goal of this technique is somewhat reprehensible in human terms but great in data processing terms. The basic idea is to create what is, in effect, the ideal totalitarian state—one in which each member has enough knowledge to do its own job but not enough to intentionally or unintentionally sabotage the system. Sounds a little like a spy novel, doesn't it? But there is no mystery about it. Some examples are in order.

In late 1985, a bank which acts as a clearinghouse for stock transactions on the New York Stock Exchange experienced an expensive lesson on the topic, "What the system elements know can hurt the system." Their system failed at a point where nearly all of the accounts which sold stock had been credited but the debits against the purchasers had not yet been run. It caused the bank to borrow more than $12 billion (as in $12,000,000,000) for a period of 18 hours. This cost them more than $5 million in interest plus the overtime and special compensation for nearly 500 programmers. Significantly enough, the failure occurred at a point where the system had processed a little over 32,000 transactions. A post-mortem showed that assumptions had been made about how many transactions might be processed in the worst case. This had become known by all programmers and used throughout the system. Since knowledge left unused is knowledge lost, literally everyone who could make use of this fact did. This is a type of pathological coupling.

Similar situations have occurred in the past, so why do they continue? One very sound reason is the fact that most of us would like to set up closed-form solutions to the data processing problems we have to deal with. We like to concoct some worst-case scenario and design to survive it. What happens is that the worst case changes unpredictably with no fanfare.

A case which we will use to demonstrate the application of the Information Hiding principle for the cause of good and against evil (being dragged out at midnight to fix a "bug" is inherently "evil") involves one of the top five insurance companies in the United States. They had a system which was responsible for processing the transactions associated with 85% of the income from a particular class of coverage they provided. Owing to a combination of factors which were never predicted by the authors of the system (only one of whom was still alive!), several changes were necessary to the format of the records the system would process. It turned out that over 400 modules would be affected (the people involved stopped counting after they reached 400 to avoid terminal depression). All of these modules used a "standard interface." What this seemingly harmless term had come to mean was that any module which needed *any* item of information from the policyholder record would read the record and use the data element(s) that were required. The problem is, this created another classic pathological couple throughout the system. This caused the structure of the record to be known by many modules. Hence, when the record format had to be changed (new fields were added, others deleted, owing to changes in federal law), these modules had to be modified. One might ask, ''Didn't they have a database management system capable of providing only a set of requested elements so that this could be avoided?'' The answer is a definite—YES! The problem is that while the vendors of such exotic systems proudly point to such capabilities, they often fail to disclose the fact that response time goes right down the drain if you use this capability. Hence, the software serves the computer and not the human.

How can we avoid this? Take a look at Figure 14.3-1. In it we have a generic-

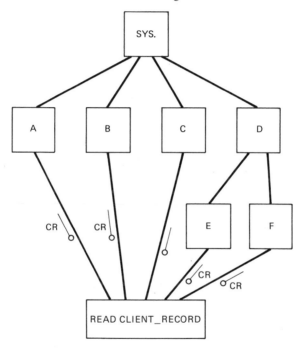

(CR = CLIENT_RECORD)

Note how modules A, B, C, E, and F will all know what the format of
CLIENT_RECORD is.

Figure 14.3-1:   Generic Structure Chart Showing Information Proliferation

looking Structure Chart. Note that there are several modules which read in the customer record. Each uses some subset of those elements to perform its individual task(s). What the Information Hiding [1] principle suggests is that we isolate or insulate knowledge of how this information is organized from the rest of the system. How can this be done? We will use the very technique that the vendors of high-priced database management systems so deftly destroyed. We will use a go-between to stand between the record and the modules which need one or more elements out of it.

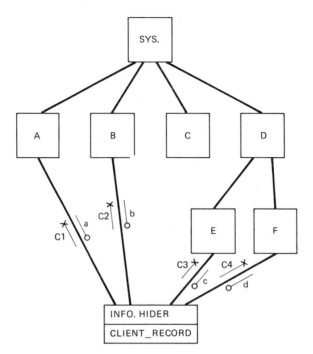

where C1 through C4 identify to module INFO. HIDER what data elements out of CLIENT_RECORD should be returned, and a, b, c, and d represent the names of different elements being returned to the modules which will use them.

Figure 14.3-2:   Revised Structure Chart Demonstrating Information Hiding

Look at Figure 14.3-2. In it we have installed a module which we have imaginatively labeled ''Information Hider.'' When a module needs one or more data elements, it calls this module, passing it a flag to indicate what elements it will need and out of what record. The Information Hiding module gets that information and returns it to the calling module. The details of how this is done may vary from language to language, but the principle is clear. The calling module does not know the physical or logical format of the information the Information Hiding module deals with. As a result, we can change the format of the record all we want, and only a single module will be impacted. The exception is the case where new data elements are to be added to support new features. In such a case, some others may be impacted, but the number will be minimal.

A word of caution: Information Hiding modules can become extremely complex. Unless we are careful, they can become a liability in disguise. In one of the worst cases of this, three Information Hiders were used. This was considered a prudent design decision because a wide variety of requests and data combinations would have to be responded to. In addition, there was concern that some sort of paging might become necessary in a future environment being considered.

We have shown one simple, common application of the use of the principle of Information Hiding. Certainly, the reader can think of many others. A sure sign that one needs to consider the use of this technique is the identification of "ripple effect" when changes to data or protocol are considered. If there is a potential for a large "ripple effect" (i.e., many effects beyond the immediate modules involved), then Information Hiding, or something very much like it, is needed.

An alternate way to utilize the concept of Information Hiding is to employ it as a means of delaying implementation decisions. For example, during the logical design of the software for an advanced type of electronic countermeasure system, the details of the hardware configuration were uncertain. The manufacturer was making some modifications that could impact system response time. The situation was unclear. Wherever the software had to interface with the outside world or pass data from one hardware processor to another, we used an Information Hider. Although there was a danger that this could become a placebo, it was done in a disciplined manner. This treated Information Hiders as "black boxes" that could get data or output data in some abstract way. The means by which each would be accomplished would be addressed during the physical design phase. This technique, used in a disciplined manner (i.e., use an Information Hider only where it is appropriate), provided several benefits:

1. It drove members of the design team, who thought we should be coding, absolutely crazy.
2. It reduced the amount of time the team spent attempting to resolve input and output issues that were unclear and capable of consuming a lot of labor hours. This helped keep us on schedule.
3. It catalogued instances of potential abuse of the Information Hiding principle. This helped immensely during physical design.
4. It helped keep the design team focused on the primary issues of logical design (i.e., architecture, meeting requirements, hierarchy, partitioning) and later, the issues of physical design (e.g., mapping to the hardware architecture, addressing constraints, performance details). This resulted in a better, more flexible design and system.

Features of the newer programming languages are capable of avoiding many of the problems associated with violations of Information Hiding. However, they may not always be used during the implementation phase unless the instances where they are needed are identified during the design phase.

14.4 QUALITATIVE EVALUATION

Have you ever wondered whether or not there is some simple, "quick and dirty" way of determining whether or not a Structure Chart you are looking at is even close to being "reasonable"? Well, there is. It is not the only technique that should be used, but it certainly fills the bill for one that is quick.

This technique is based on the fact that the partitioning we so painstakingly arrived at during the analysis phase and tried to maintain during the design phase resulted in the identification of functions. These functions were not arrived at in the usual way by listing tasks or flowcharting but by examining dataflow and data transformations. Given that we arrived at this point in time by this route, we can use this method without getting into too much trouble. It is based on a concentration of all of the observations, advisories, guidelines, and "good things to do" that we have discussed. It also takes the effects of prudent coupling and cohesion and puts them into practice.

Table 14.2-1 presents these guidelines in a compact form. Note that we are basically applying the same sort of analysis to a Structure Chart that a heat-transfer engineer might apply in solving a heat-transfer problem. We are dealing with the extremes (also known as boundary conditions) and using those to determine what is going on inside the chart. For example, the bottom of the chart is relegated to the processing of "dirty" or potentially invalid data, while the top deals only with logicalized data. Things get somewhat muddled in between. Also, the volume of data is high at the bottom and very low at the top of the chart. If you are looking for a simple way to evaluate your work or that of others, use this chart as a starting point.

TABLE 14.4-1: Quality Guidelines for Structure Charts

Location in Structure Chart	Shape	Data	Type of Information	Type of Activity	Module Reusability
Top	Narrow	Low-volume, logicalized	Control	All management	Low
Middle	Broad	Mixed (clean and "dirty")	Mixed (work and management)	Mixed (work and management)	Moderate
Bottom	Somewhat narrow	Unvalidated (dirty)	Data	All work	High (if well factored)

14.5 FUNCTIONAL ANALYSIS

This approach applies two principles: inclusion and decomposition. We used the concept of decomposition in breaking processes and dataflows down from large aggregates into smaller and smaller pieces. Each decomposition required that we carefully determine that we had not gained or lost any information along the way. But inclusion involves something a little different. By inclusion we mean that a concept or task or function logically includes other supportive or less formidable concepts, tasks, or functions. As a result, a physical object can be broken down. For example, an automobile can be decomposed in any of several ways as shown below:

$$automobile = body + chassis + propulsion\_system + frame$$

or

$$automobile = electronic + hydraulic + mechanical + thermodynamic$$
$$+ electromechanical$$

or

$$automobile = ferrous\_metals + nonferrous\_metals + petrochemicals$$
$$+ amorphous\_solids + electrolytes$$

Certainly, there are many other ways in which we could break down the automobile. Which is the "right" way? It depends on what we wish to do with the breakdown. In our case, the breakdown that we have thanks to Structured Analysis and some earlier work to refine those results is based upon just what we are looking for—information transformation. Hence, we should be able to check out whether or not the Structure Chart is any good on a gross level by looking at it on a level-by-level basis. We will use the Structure Chart in Figure 14.5-1 for our discussion.

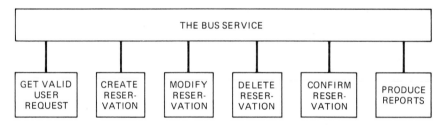

Figure 14.5-1:   Structure Chart Demonstrating Functionality

In examining the figure, note that each boss module has immediate subordinates which are directed at accomplishing the task the boss is responsible for. Hence, we should be able to read the title of the boss (e.g., Bumble's Bus Service) followed by the names of the modules which are its immediate subordinates in order to find out how the task is accomplished. The process is as follows:

```
FOR EACH BOSS MODULE, DO

     Form a sentence with the name of the boss (N) and the names of its
     immediate subordinates (S1 through Sm) of the form:

     ''The N system — S1, S2, S3,  . . . , and Sm''

ENDDO
```

For example, in the case of the bus service problem that we looked at earlier, this approach would result in a sentence of the above form that read:

```
The BUS SERVICE SYSTEM  —  GETS USER REQUESTS, MODIFIES RESERVATIONS,
                           CREATES RESERVATIONS, DELETES RESERVATIONS,
                           CONFIRMS RESERVATIONS, AND PRODUCES REPORTS
```

The key to determining whether or not the partitioning or design structure is even acceptable is whether or not the sentence which results from this process seems reasonable. This reasonableness test is applied in two ways:

Appropriateness.   This refers to whether or not the level of detail and the process-oriented nature of the sentence is consistent with our understanding of the software design. For example, finding some low-level task at the top level in the system indicates that the partitioning and/or hierarchy are inappropriate.

Completeness.   If the design is complete, then not only should the level of detail of the modules at the highest level be consistent, but the design should be complete. Completeness means that literally everything that one could think of that the system will have to do "belongs" logically to one of the subordinate modules reporting to the boss. This is analogous to having each module that reports to the boss play the role of a "pigeonhole." Each and every task, however low-level, must fit into one of these pigeonholes. If we can think of one which does not appear to have an appropriate slot, some rework may be in order. If it is not clear which slot a given low-level task belongs to, we may or may not have a problem. Further investigation will be necessary to determine this. In our example, no "pigeonhole" exists for accounting for funds.

14.6 RELATIONALITY

Most of us have heard that the structure of a solution should emulate the structure of a problem. In data processing, we take this to mean that there should be a strong relationship between the functionality of a system (the solution) and the functional need (the requirements). More recently, some valid concern has arisen over the mapping of the information to the system functions and hierarchy. For example, the "do everything" module may have logical cohesion. As a result, it may also be capable of processing a wide range of data. We are referring not just to the values of the data being processed but also to the entities that the individual data items are a part of. It has been observed that modules which are not very cohesive tend to process data from different entities. Hence, the quality of a module and of the system it is part of depends to some extent upon the degree to which different entities are processed by different modular families.

The principle of relationality may be stated as follows:

> The quality of a software design is partly a function of the degree to which strongly related sets of data are processed by strongly related modules within the system.

An alternative way of stating this principle is that the structure of the software should emulate the structure of the data. Following this assumption, the relationships that are present in the data would be somewhat paralleled in the software design (and the system itself). This is similar in concept to the Warnier-Orr method [2, 3] discussed in Chapter 10. However, it embodies one seemingly insignificant but important difference. The initial basis for the analysis and the design is the *flow* of data through the system. Only coincidentally does the data structure get involved. The side effects of the dataflow orientation are interesting in that one of them is to create software designs that meet the data-structure-based criteria. The reason for the preference of *dataflow* is its flexibility and the ease with which one can communicate complex concepts such as real-time interactions.

REFERENCES

1. "On the Criteria to Be Used in Decomposing Systems into Modules," by D. L. Parnas, originally appeared in *Communications of the ACM,* Vol. 5, No. 12 (December 1972), pp. 1053–58. Copyright © 1972 by the Association for Computing Machinery, Inc.
2. J. D. Warnier, *Logical Construction of Programs,* 3d ed., trans. B. Flanagan. New York: Van Nostrand-Rheinhold, 1976.
3. K. T. Orr, *Structured Systems Development.* New York: Yourdon Press, 1977.

CHAPTER 15

Structured Design
as a Process

Structured Design proceeds through a series of stages as we move from an initial cut at the design (see Chapter 10) to an actual "blueprint" of the system to be built. The major phases that design of any sort proceeds through are the Logical or Conceptual Design phase and the Physical or Detailed Design phase. That is the order in which they occur. First we define a concept (or concepts) upon which the design will be based, then we proceed to embellish and refine that concept, modify it to accommodate the environment the system will be operating in, then finalize the design in the form of an as-built or to-be-built blueprint.

In a manner similar to Structured Analysis, the activities in Structured Design can go on interminably. It is not clear that the original form of this method contained any sort of "stopping rules" which enabled us to know whether we should proceed or quit. This lack has led to problems within many projects. It causes a lack of focus on specific tasks and predictably reduces productivity. An important facet of Structured Design that we will dwell on in this chapter is the creation of a sequential task list which is tailored for use with Structured Design. Such a list may not be unique, but this one incorporates specific deliverables, associated with certain sets of tasks which occur throughout the software design activity. This mechanism tends to focus effort on the accomplishment of specific, tangible goals. Such focusing has been shown by many management and business studies to improve productivity, since each individual knows what is expected and how the results will be assessed. We refer to this task list as a Product Oriented Work Breakdown Structure.

We will describe the process of developing a Structured Design in light of a series of activities directed at the production of a series of products. Notice that no line item in this task list describes an activity such as "Document Design." This is because a truly effective *system* for software analysis and design and implementation would have documentation occur as an incidental part of doing the work, *not* as some sort of afterthought.

15.1 THE PHASES OF DESIGN

Before discussing the development of a Structured Design, we need to understand the generalized process that we are going through. Design activity has been studied from many viewpoints. Although there are many different models of the process, these can be summarized into three primary stages:

Initiation. This is the form the design takes when it has first been generated. It is usually not complete, in that some functions overlooked in the analysis may still be missing, error processing is missing, and details of how outputs are produced (particularly reports) will also be missing.

Abstraction. Derived directly from the initiation results, this design phase involves the completion of the design. All functions and necessary detail missing in the abstraction phase are incorporated into the design. At this point, no consideration has consciously been made to accommodate the specific characteristics of the hardware, target programming language, or timing and sizing issues. This is the phase in which we apply coupling and cohesion, obtaining the best design we can given the constraints of time and talent.

Instantiation. Based on the abstract design, this phase involves the revision of the design to make use of specialized operating system features, the use of hardware features rather than software to provide needed services, and the incorporation of any additional functions or features that have been identified as being needed. During this stage the software engineer makes rational compromises with respect to coupling and cohesion in order to attain some performance goal or to meet some external constraint.

Structured Design organizes the above stages into two phases:

Logical Design. In this phase, the objective is the development of a design which is directed at an abstract machine and operating environment. This is analogous to the development of an artist's conceptual drawing prior to the development of the blueprint of a physical system (e.g., an automobile, a shopping center). This model contains the highest amount of quality and quality components as we were able to incorporate, given the time and talent available. Little or no conscious consideration has been given to the implementation issues.

Physical Design. The objective of this phase is the creation of a blueprint of the system. This blueprint is complete and correct and accurately describes *every* aspect of the system to be developed. Literally, the as-built system will look exactly like this blueprint. Just as in the construction industry, there may be instances when the construction of the software reveals that a departure from the blueprint is advisable. So be it. However, the blueprint will be changed if for no other reason than posterity ought to know. Besides, those who have to maintain the system (who may be the developers themselves if they are not careful) will need some sort of map of what the system is like. Another way to look at this is that if the system's blueprint either does not exist or is so inaccurate that it practically does not exist, then those who developed it are going to drop into something of a career "potential well." That is, their participation will be required in every change and modification of the system. Some consider this to be job security, but it is, in effect, a death warrant to future marketability within the company and without. How many companies are going be bid competitively for someone whose claim to fame is that they have been working on the same bill of materials system for the last 14 years?

15.2 DISTRIBUTION OF RESOURCES WITHIN THE DESIGN PHASE

A question often asked is how many person-hours will be required to accomplish a particular design. The exact number is definitely something of a "guesstimate." However, the distribution of the hours allocated to the design activity is not. Specifically, the advised percent distribution between analysis, design, and implementation is 50%–60% for the analysis and design phases combined and 40%–50% for the implementation phase. Within each of these phases a further breakdown is advisable. A set of advisory guidelines of this type are presented in Table 15.2-1. These are only advisories and should be tailored to each project.

TABLE 15.2-1: Percent Resource Distribution by Phase

Phase	Subphase		Percent Allocated
Analysis			35%
	Current Physical	20%	
	Current Logical	5%	
	New Logical	10%	
Design			25%
	Logical Design	20%	
	Physical Design	5%	
Implementation			40%
			100%

15.3 WORK BREAKDOWN STRUCTURE FOR THE DESIGN PHASE

As described earlier, our model of Structured Design holds that it is composed of two distinct but related phases. In chronological order, the first is the Logical Design phase and the second the Physical Design phase. Each of these is described below in terms of the activities and the products associated with sets of activities.

15.3.1 Logical Design Phase

The activities which occur during the Logical Design (also referred to as Preliminary Design) phase are described in Figure 15.3.1-1. The activities listed there are further described in Table 15.3.1.1.

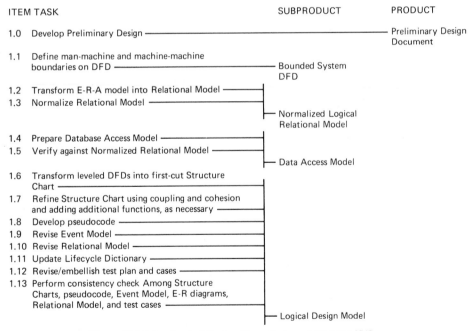

Figure 15.3.1-1: Logical Design Phase Tasks and Products [S1]

TABLE 15.3.1-1:   Description of Logical Design Phase Tasks

1.0 Develop Preliminary Design
     The goal of this task is to create a conceptual design which will reflect the requirements spelled out
     in the analysis package and contain the maximum amount of quality that we can incorporate into it.
     In this context quality is synonymous with maintainability and reduced development cost.

1.1 Define man-machine and machine-machine boundaries on DFD
     Using the level-'0' dataflow diagram developed during the logical modeling phase of analysis,
     identify which processes or functions will be allocated to hardware, software, and manual
     procedures.

1.2 Transform E-R-A Model into Relational Model
     Each entity and each relation is transformed into a logical relationship using the Relational
     Database Model notation and semantics. The relations are integrated into a global relational
     schema.

1.3 Normalize Relational Model
     The logical relational schema is transformed into a normalized schema to provide for logical data
     access and update integrity.

1.4 Prepare Database Access Model
     For each user view a logical data access specification is created, which defines the process followed
     in accessing the required data and the logical relations involved.

1.5 Verify against Normalized Relational Model
     Inability to support required data access will cause the E-R-A Model and the Logical Relational
     Model to be refined appropriately.

1.6 Transform leveled DFDs into first-cut structure chart
     Using a translation process, create the structured design structure chart which corresponds to the
     dataflow model developed during the new logical modeling process in analysis. This is strictly a
     first-cut design and will need further refinement. Each process in the DFD at level '0' that is
     decomposed to a lower level must undergo the same translation process to complete the first-cut
     Structure Chart prior to beginning the refinement process.

1.7 Refine Structure Chart using coupling and cohesion and adding additional functions, as necessary
     The first-cut Structure Chart does not include the read, write, and user interface functions that were
     left out of the logical form of the analysis. These must be added to the Structure Chart. In addition,
     many of the features and processes that resulted from the analysis are being accomplished in a
     manner which is analytically acceptable but does not constitute ''prudent'' design practice, even
     for a logical design. These can be identified by applying the principles of coupling and cohesion as
     well as (later) additional software design heuristics.

1.8 Develop pseudocode
     The pseudocode for each module describes what the module does. The basis for most pseudocode
     is the structured English that was created during the analysis phase. If pseudocode was utilized
     during that phase, then it needs to be reviewed/revised in order to meet the more rigorous demands
     of the design effort.

1.9 Revise Event Model
     Refinement of this model is necessary in order to put it into a more acceptable form to accommo-
     date the design and implementation perspectives.

1.10 Revise Relational Model
     Bring this view of the system into compliance with the other revisions and refinements that have
     taken place.

1.11 Update Lifecycle Dictionary
     One of the outcomes of transforming the analysis results into an initial design model is that new
     data elements have to be incorporated. These are flags or control couples which are used by
     modules to signal other modules of the existence of certain conditions. The status of these condi-
     tions will affect the way in which the system executes. Specifically, the modules which execute and
     the order in which they execute will be affected. Since the Lifecycle Dictionary is the repository for
     such information, it must be updated to reflect the creation of these new information elements that
     are part of the system. In addition, the Lifecycle Dictionary will also contain descriptions of each of
     the modules in the system, the test cases to be applied to them, and their cohesion level. The
     coupling level between modules is included in the dictionary under ''system description.''

TABLE 15.3.1-1: (continued)

1.12 Revise/embellish test plan and cases
When the test plan and accompanying test cases were first created, the details of what the system was to do were not detailed. As we have proceeded further into the design phase, we have identified several kinds of tests that would be appropriate for the system to perform in order to ensure that it can continue to function in an acceptable manner despite problems that may be encountered. A review, revision, and refinement of the content of the test plan is in order at this point to capture this new knowledge.

1.13 Perform consistency check among Structure Charts, pseudocode, Event Model, E-R diagrams, Relational Model, and test cases
Ensure that all parts of the design are accurate and mutually supportive.

15.3.2 Physical Design Phase

The Physical Design phase is described in a manner similar to the Logical Design in Figure 15.3.2-1 and Table 15.3.2-1.

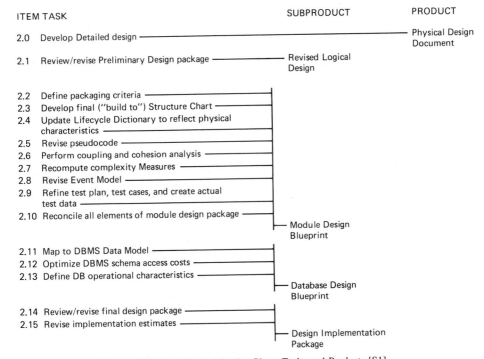

Figure 15.3.2-1:   Physical Design Phase Tasks and Products [S1]

TABLE 15.3.2-1:   Description of Physical Design Phase Tasks

2.0 Develop Detailed Design
The Detailed or Physical Design is intended to be a blueprint of the system to be built. It is derived from the Logical Design by incorporating considerations of system size, timing constraints, the existence of hardware or software utilities, language features, and other implementation considerations. Literally everything about the system to be built is described and documented during this phase in order to simplify and expedite the implementation effort.

2.1 Review/revise Preliminary Design package
Even though it has been walked through, it is necessary to make a "final" pass over the contents of the Logical Design. This is particularly important if the group who will be doing the implementation and/or the detail design of the system are not the same group of people who developed the logical design. The purpose of this task is twofold. One is to identify and correct errors and necessary refinements. The other is to ensure that all that is necessary to accomplish the physicalization of the design is present and of sufficient quality to support this next phase.

2.2 Define packaging criteria
The term "packaging" is applied to the process of incorporating implementation considerations into the Logical Design. This cannot be done without some basis or value system being applied. The problem is that everyone involved will attempt to apply their own, such as the importance of using or avoiding certain language or operating system features. This necessitates the creation and acceptance of a common set of guidelines upon which the packaging decisions will be based. In essence, this amounts to incorporating everyone's opinions into a mutually agreed-to compromise position.

2.3 Develop final ("build to") Structure Chart
This amounts to the application of the packaging criteria in such a way that there is minimal degradation, if any, of the various quality characteristics of the system. For example, in some cases it may be more prudent to utilize common coupling, owing to the nature of the hardware and software environment the system will be operating in. However, this does not mean that common coupling is the *only* means by which data will be exchanged. Only a case-by-case decision-making process will effectively retain as much quality and coherence as possible in the final system.

2.4 Update Lifecycle Dictionary to reflect physical characteristics
The Lifecycle Dictionary must include the compromises between the desired (Detailed Design) system and the obtainable (Physical Design) one. The primary inputs or changes at this stage are: the "build to" Structure Chart process, dataflow and data access specifications; the implementation constraints from the packaging criteria; the changes to the Event Model; and the new DBMS Data Model and its operational characteristics.

2.5 Revise pseudocode
Update the contents of the pseudocode, as necessary, in order to accommodate the changes made to the system design during the packaging process.

2.6 Perform coupling and cohesion analysis
Owing to the nature of the software and hardware environments available today, we need to make sure that what we finally end up with for a blueprint is justifiable. That is, no design or programming practice is totally without merit, but in each instance it must be justified on its own merits. The basic question that we are attempting to answer is, "Given the other alternatives we had available, are we satisfied that this is the best?"

2.7 Recompute complexity measures
It is quite possible that, during the process of physicalizing the Logical Design, nuances have been incorporated which have made some modules more complex than they were before. The recalculation of complexity factors will help to establish whether or not a reexamination of some modules is warranted.

2.8 Revise Event Model
This may be necessitated by "discoveries" or changes in approach occurring during packaging.

2.9 Refine test plan, test cases, and create actual test data
This activity brings us to the point of having as complete a test package as we can prior to the implementation of the system.

2.10 Reconcile all elements of module design package
As before, this is something of a quality assurance activity.

2.11 Map to DBMS Data Model
The logical, normalized relational schema is translated into an initial DBMS schema (hierarchical, network, or relational). This schema is then refined to take advantage of the unique capabilities of the DBMS being used. The translation process is specific to the particular DBMS Data Model. Each translation starts with a logical relation and converts it into an appropriate DBMS structure.

TABLE 15.3.2-1:   (continued)

2.12 Optimize DBMS schema access costs
       Cost models are used to determine optimal access paths to support the user data view requirements.
       Any modifications to the DBMS schema are verified against the logical relational schema.

2.13 Define DB operational characteristics
       These characteristics typically include:

         batch, online transaction distributions
         database availability and integrity
         database security and privacy constraints
         operational reporting requirements

2.14 Review/revise final design package

2.15 Revise implementation estimates
       This supports the concepts of iterative estimation refinement that is used in most engineering fields.

15.4 WHERE ADVANCED STRUCTURED ANALYSIS AND STRUCTURED DESIGN ARE HEADED

In the previous chapters we have examined an updated form of Structured Analysis and Structured Design, explained their use, and demonstrated their application to specific problems. We have repeatedly noted that these two methods represent a real breakthrough in software specification and design technology. They have repeatedly demonstrated their effectiveness on a broad spectrum of software system applications.Together, they provide an inexpensive and effective way to more coherently model client problems, communicate with clients, and ensure that the design reflects exactly what is needed—not something else. They have also shown themselves to be resilient in the face of change and flexible enough to respond positively to it.

Perhaps their greatest contribution is the fact that they provide us with a means of addressing all three aspects of *any* software system. Based on experience, addressing control, dataflow, and information relationships separately has been a good one. This strategy has enabled us to break complex systems problems down into simple ones, solve those, and maintain the relationships among all elements. The key to all of this is the concept of a Lifecycle Dictionary. Its name is misleading. Project resource might be a better name. It's automation and support have been a primary contributor to the success of several projects. As noted earlier, its enhancement and automation has begun and will continue. As we look to the future for these methods and their successors, an increase in the use of the systems view described here will make these conceptual tools even more effective.

REFERENCES

1. M. Page-Jones, *The Practical Guide to Structured Systems Design*. New York: Yourdon Press, 1980.

2. E. Yourdon and L. L. Constantine, *Structured Design.* New York: Yourdon Press, 1978.

S1. Extracted from the seminar, "Structured Design of Real-Time Systems," by Software Consultants International, Ltd., Kent, Washington; Copyright 1988. Reprinted with permission.

Guidelines for Informal Walkthroughs

The concept of walkthroughs may have first been publicized in 1971 in the book, *The Psychology of Computer Programming* (G. Weinberg). Although the concept seemed fresh and beneficial, it did not exactly set the computing world back on its heels. The main purpose of walkthroughs at that time was to identify and correct program errors and suggest refinements. An additional and important side effect was that more than one individual in an organization would become familiar with a program. This reduced the risk borne by the company which paid to have the software developed.

More recently, the use of Structured Analysis, Structured Design, and Information Modeling has underlined the fact that errors can occur in more than computer programs. They can occur in both analysis and design. More important, views may differ on how a particular data item should be defined. These differences of "opinion" can be significant. Many organizations have found it almost essential to conduct walkthroughs on each and every element of a product of the analysis and design phases. This practice, like the practice of walkthroughs of code—and for the very same reasons—has proven invaluable.

This appendix summarizes a set of guidelines which have proven to be effective in walkthroughs of analysis and design material. These are based on the experiences of several organizations who have had a great deal of success with their use. They are intended as a starting point for an organization to develop its own forms and standards for use with walkthroughs.

Walkthroughs are of two types:

Formal. These are usually required under some contractual obligation or by internal corporate policy. Such walkthroughs are usually defined by contract as to their content, format, and conduct. They are *not* the type of walkthrough that we will be discussing in the rest of this section.

Informal. These might also be called "working walkthroughs," because they occur during the lifecycle as the work is being performed. Whereas there may be only a few formal walkthroughs, there may be an undefined number of informal walkthroughs.

A.1 POLICIES

No analysis or design is considered complete until it has successfully been walked through informally *without* the identification of errors, interface problems, or incomplete information.

The results of each and every informal walkthrough will be documented in a project notebook maintained by the project librarian, lead analyst/designer, or secretary.

The project notebook will be divided into three sections. One section will contain material that has been developed but not walked through. Anothe will contain material that has been developed but not walked through. Another will contain material which has been developed and walked through but is in need of further refinement. A third section will contain material which has been developed, walked through, and corrected in response to walkthrough

Any change to material which has been walked through without remark (error) will cause it to be reclassified and walked through again.

The author of the material to be reviewed may change any of it prior to the meeting. However, this revised material must be handed out as soon as it is available.

Supervisors may not attend informal walkthroughs. They may be briefed in meetings wherein there will be no criticism of work by colleagues.

Attendees are to spend a *minimum* of one hour reviewing the material to be discussed at the walkthrough.

Anyone who has responsibility for any part of the system interfacing with the material being reviewed should be included in the attendees list.

The person(s) who called the review are responsible for explaining the materials and answering questions.

As a *minimum,* review materials for analysis must include dataflow diagrams (no more than two levels may be reviewed at one time), corresponding dictionary entries, and pseudocode or structured English. Review materials for design must include the Structure Chart, expansion of that portion being reviewed, dictionary entries, and pseudocode.

A.2 SCHEDULING

An informal walkthrough may be scheduled by any analyst or designer through the person whom that analyst or designer has designated to be the coordinator of the walkthrough.

Attendees for an informal walkthrough will be notified *at least* 24 hours in advance of the scheduled walkthrough time.

Walkthrough schedule will include *both a start and an end time*. It is the responsibility of the author of the material being reviewed and the review coordinator to see that the schedule is maintained. If the review results in the identification of issues which cannot be resolved at the review, the coordinator is to assign responsibility and a schedule for their resolution. They are not to be resolved in a group mode.

One person will be designated by the moderator to take notes during the review. These notes will include the identification of the attendees, comments/corrections to the materials, and schedule for the correction of identified shortcomings of the material.

A follow-up review will be scheduled, if at all possible, before the review is adjourned.

Automated Support
for the Structured Methods

In the last year several software systems have been produced and heavily marketed which are purported to support use of the structured methods. Most have been touted as ''productivity enhancements.'' That is, they will improve the productivity of the persons using them. For purposes of this discussion, we will set aside the question of whether or not productivity is measurable or even appropriate to apply to any profession, whether it be engineering or software development. Instead, we will focus our attention on the tools themselves.

For those of you who have not developed a software product for public sale and distribution, a few givens must be related. Most product developments of the type that yielded the products now available to support the structured methods were financed by ''venture capital.'' In this arrangement, a group of investors supplies the funds for the product to be developed, packaged for distribution, and marketed. Usually, the marketing and packaging of the product will account for most (two-thirds to three-quarters) of the funds used. As you might expect, this is a risky gamble for everyone involved. The venture capital investor usually wants to get the initial investment back in as little as two years and see about a tenfold (that's right folks, 10×) return on investment in the first five years the product is out. People who develop such tools are thus under great pressure to get something out as soon as possible and to produce what can be marketed immediately. All of this has lead to a shameful situation with respect to automated aids for the Structured Methods.

One's main concern when considering the purchase of a software tool of this sort *should* be whether or not it will be able to do some of the drudgery work that must now be done manually. Examples of such drudgery are manually creating and verifying the dictionary. For example, it is a labor-intensive activity to establish that every item mentioned in the pseudocode (or structured English, if you prefer) is defined in the dictionary. It is also drudgery to determine whether or not everything in the dictionary is actually referred to somewhere in the pseudocode. At this writing, *none* of the software tools available does this! What, then, do they do?

Surveys of such tools reveal that the emphasis universally has been on the graphics—that is, the ability to draw and maintain dataflow diagrams and structure charts. This is great for sales—but, as someone once said, a dataflow diagram without a dictionary is "just a pretty picture." In fact, the ability to level and balance dataflow diagrams was a "missing" feature that only recently was added to the most popular of these tools. Unfortunately, this balancing is done only on a level-by-level basis, which allows overall imbalances to occur. In addition, none of these tools can cross-check the interrelationships between/among the various aspects of the system analysis or design—that is, make sure that the process model is consistent with the event and information models and conversely.

So just what do these tools do? They draw pictures. Some are better at it than others, some are more convenient than others, and one in particular has a nasty habit of "losing" your diagrams if you change then in just the right way (none of this is mentioned in the enormous user's guide that comes with it).

Where does all this leave you, the practitioner? Quite simply, you would be better off to wait on that purchase. Right now, it is not clear which system is best or which company will survive financially. What *is* clear, right now, is that one would be better off with a high-quality, technically sound dictionary tool—one that did all we mentioned above and then some—linked to some utilitarian graphics or plotting package. It would cost a lot less and would reduce frustration with packages that were simply not developed by people who have actually used these methods *on real systems*. This last point is a sore one. When you encounter a vendor with one of these tools, they rarely let anyone use it, nor do they let you enter into it anything approaching the kind of system where tangible benefits of such a tool would be cost justifiable.

Index